To my dearest Mom and Dad,

Your steady backbone has been my guiding light during the vast expanse of the sky, where galaxies dance in perfect melody and stars glitter like dreams. This book, **Marvels Of The Cosmos**, is an expression of your wisdom, love, and support for me.

My curiosity has been sparked by your caring nature and eternal affection, Mom, in the same way that the sun feeds the energy that powers our universe. Dad, you have been my cosmic constants—your courage and resolve keeping me grounded in the face of adversity like a celestial force holding galaxies together.

I get a glimpse of the wonder that is our family as I delve deeper into the mysteries of the cosmos on these pages. The way stars create constellations, so too has your presence moulded mine; every passing moment is a beautiful star that adds to the fabric of our journey together.

With deep gratitude and immense love,

Pabitra Banerjee

01/01/2024

Marvels Of The Cosmos

Pabitra Banerjee

Contents

Introduction

Conveyors

Galactic Dust

Collapsing

Fortune

About The Author

Copyright

Introduction

The Universe, as revealed by modern research, is dazzlingly beautiful and devastating in magnitude. It is 45.6 billion light years vast, 13.7 billion years old, and contains 100 billion galaxies, each of which contains hundreds of billions of stars. However, surprisingly, the gap between us and the cosmos has vanished as our understanding of it has grown. Even if the universe turns out to be infinitely large and full of unimaginably strange worlds, modern science says that we require it all to survive. There wouldn't be enough resources to construct humankind without the stars, and the vast antiquity of the universe wouldn't allow the stars to complete their magical work. The universe cannot be old without being large; if there are to be spectators to marvel at its glories, there may not be any waste or redundancy in this potentially endless domain.

Thus, the history of the universe is also our history, taking us back beyond the beginning of humankind, the emergence of life on Earth, and even the formation of the planet itself, to possibly inevitable or coincidental events that happened less than a billionth of a second after the universe first came into being.

✦ A HISTORIC GEM

During the peaceful atmosphere of Christmas Eve 1968, Apollo 8 smoothly climbed into the Moon's shadow. Frank Borman, Jim Lovell, and William Anders became the first humans to see Earth disappear from view while they were inside that heavenly prison. A stunning sight greeted them as they emerged from the lunar shadow: a delicate crescent Earth rising against the vast background of space. They decided, in a profound moment, to tell the inhabitants of their beloved home planet a cosmic genesis story. Pilot of the lunar module William Anders, five hundred thousand miles from home, began:

"We are currently nearing lunar sunrise, and, for all those on Earth, the Apollo 8 crew has a message we wish to convey to you.

In the beginning, God created the heavens and the Earth.

And the Earth was without form, and void, and darkness was upon the face of the deep. And the Spirit of God moved upon the face of the waters.

And God said, Let there be light: and there was light.

And God saw the light, that it was good, and God divided the light from the darkness."

A common theme in many cultures' origin myths is the rise of light from darkness. The Maori people called the beginning of the universe **Te Kore**, while the Greeks called it **Chaos**. According to Egyptian belief, the earth and the gods came from an endless, unfathomable ocean that existed before the birth of the universe. Certain civilizations believe that God is everlasting, having created the universe from nothing and existing long after it. In others, like certain Hindu faiths, the Earth and the skies are older than a massive primordial ocean. Only when light emerged and the darkness was driven away did Lord Vishnu awaken and give the order to create the world while he floated on the ocean, ensnared in the coils of a great cobra.

Though the exact origin of the universe is still unknown, there is compelling evidence that something remarkable occurred 13.8 billion years ago, which may serve as the universe's creation. It's known as the Big Bang. (We must take caution when choosing terms here, as this is a book on science and the ability to differentiate between the known and the unknown is essential to excellent science.) This intriguing event aligns with the beginning of everything that is currently visible in the sky. A volume much smaller than an atom once had all the components needed to create hundreds of billions of galaxies and thousands of trillions of suns.

This little seed, which is unbelievably dense and hot, has been expanding and contracting for the last 13.8 billion years. During that time, the laws of nature have had enough time to construct all the complexity and beauty we see in the night skies. In addition, Earth, life, and consciousness were created by these natural processes. These concepts are often more difficult to understand than the origin of the seemingly endless stars.

It is important to exercise caution since our current understanding of the early universe, which we define as the events that occurred during the Planck era, or the time period before a million million million million million millionths of a second after the Big Bang, is currently beyond reach. This is due to the fact that, prior to this time, we had very little to say about space and time as we lacked a theory of it.

The greatest mystery of modern theoretical physics is a theory like this one called quantum gravity, which has hundreds of experts worldwide searching in earnest for it. (Albert Einstein looked for it in vain for the last many decades of his life.) According to conventional wisdom, time and space both started at time zero, or the start of the

Planck period. Therefore, the Big Bang might be considered the beginning of the universe, marking the beginning of time itself.

But there are other options. According to one explanation, the collision of two **branes**, or fragments of space and time, that had been floating in an endless, pre-existing space for all eternity, resulted in what we see as the Big Bang and the beginning of the Universe. Therefore, nothing more substantial than a cosmic collision of two separate layers of space and time was what we identified as the beginning.

The solution to the age-old question, "Why is there a universe?" may never come to pass, or it may be discovered during our lifetimes. Regardless, the journey has proven to be more worthwhile than the result because the investigation into the universe's origins is at the core of science. In fact, it forms the foundation of a large portion of the evolution of human culture. The great civilizations of antiquity all developed beginnings, origins, and endings to their myths, suggesting that the desire to comprehend events beyond Earth is natural. We have only lately learned how extremely beneficial this search is from a practical standpoint. In addition to improving our understanding of nature, this search has given us the ability to manage and control it through technology, improving our quality of life. This is possible when the scientific method is applied. Curiosity is the foundation of everything we take for granted, from cross-continental flight to medical advancements.

✚ THE WORTH OF MAGIC

At the base of **Marvels Of The Cosmos**, is the notion that a trip to the edge of the universe has profound implications for our daily lives. I can't stress enough how much I believe that the basis of civilization is intellectual and physical exploration. To use the words of my native Manbazar, "constructing rockets to the Moon and crafting telescopes to capture the light from the farthest stars may appear as an intriguing luxury," but such a vision would be superficial, inaccurate, and downright foolish. We exist in the universe and it lives in us; we are a part of it, and its fate is also our fate. What could be more significant, applicable, and helpful than comprehending how it functions?

As I started to plan this, I wanted to create something that went beyond a brief introduction to the majesty of the cosmos. Granted that we witness black holes, crashing galaxies, and stars at the edge of time, it would be inaccurate to dismiss the old study of astronomy as a spectator sport. Through the lens of our telescopes, the wonders we witness serve as laboratories in which we can test our understanding of the natural world under conditions so harsh that they will never be repeated on Earth. In light of this, i chose to centre the events around scientific concepts as opposed to the wonders in itself.

Light is the only thing that connects us to the far-off Universe, which may always be out of reach. This is the theme of the narrative **Conveyors**. The most distant starlight contains the chemical fingerprints of the elements, which allows us to determine with absolute certainty the makeup of the star. However, it is also about the information contained in light itself and how that information got there.

What are the universe's building blocks? is an age-old question posed in **Galactic Dust**. Furthermore, from the remnants of the Big Bang— a scorching hot, exquisitely structured fireball devoid of any recognizable structure—how was the basic building block of the human being assembled?

Collapsing narrates the tale of gravity, the universe's master sculptor. Gravity is by far the weakest of the four fundamental forces in the universe for reasons that are beyond our comprehension, but it has a limitless range and operates upon everything, thus its impact is all-pervasive.

The most precise theory of gravity is Einstein's General Theory of Relativity, published in 1915, and it remains our most accurate explanation of gravity. The concept of the Strong Nuclear Force emerged in the 1960s and 1970s. Quantum Electrodynamics, the theory explaining the electromagnetic force, originated in the 1950s. The Weak Nuclear Force is described by the Standard Model of particle physics, a theory dating back to the 1970s that combines quantum electrodynamics with the explanation of the weak nuclear force. It is important to note that the Higgs Boson, a crucial component, has already been discovered. This discovery was made at CERN's Large Hadron Collider in Geneva, providing a comprehensive understanding of the Weak

Nuclear Force and its relation to electromagnetism.

Nevertheless, despite the long pedigree and the exquisite accuracy and elegance of Einstein's theory of gravity, it is acknowledged to be incomplete. Our understanding of the universe encounters limitations in the heart of its most intriguingly named phenomena. Black holes, recognized to exist at the centre of galaxies like the Milky Way, are scattered throughout the cosmos, representing the remnants of the most massive stars in the universe. Their presence is discerned through their influence on neighbouring stars and the detection of intense radiation emitted by gas and dust unfortunate enough to approach their event horizons. The formation of black holes has been observed in the aftermath of the most violent cosmic events, such as supernova explosions. These occurrences signify the demise of stars that once illuminated the cosmos for millennia, now culminating in a matter of minutes.

The concluding chapter, titled **Fortune**, delves into the distant past and the far future, tracing the inexorable progression of the universal clock. It is within this chapter that the profound contribution of engineering directly intersects with our narrative. The science of thermodynamics, which now serves as our compass for understanding the ultimate fate of the Universe, originated from considerations of the efficiency of steam engines in the nineteenth century, rather than a deliberate quest to peer into a potentially infinite future.

In **Fortune,** we intricately explore thermodynamics, illustrating how this quintessentially nineteenth-century science provides us with a tangible foundation to speculate about events that will unfold countless years from now — a staggering figure stretching into unfathomable lengths: 10,000,000,000,000,000,000,000,000,000,00 0,000,000,000,000,000,000,000,000,000,000,000 ,000,000,000,000,000,000,000,000,000,000,000 years. This remarkable foresight, born from the pioneers of the age of steam, demonstrates the enduring relevance and incredible reach of their scientific endeavours.

As we cast our gaze toward the future and explore the marvels of our universe, we unearth a fascinating revelation within Einstein's theory of gravity, our most refined depiction of the fabric of the cosmos. This theory foretells the demise of the universe within black holes, where the collapsing remnants of the most brilliant stars serve as the frontier of our comprehension of the laws of physics and, consequently, the outer limits of our grasp on the wonders of the universe. This juncture, at the precipice of the known and the unknown, is precisely where every scientist aspires to stand.

The term **science** carries manifold meanings; one might define it as the collective knowledge constituting the grand library of the known. However, the true essence of scientific practice unfolds at the border between the known and the unknown. With a foundation built on the contributions of giants who came before, scientists peer into the darkness not with trepidation but with wonder. The fervent aspiration of every scientist is not only to catch a glimpse of something necessitating a new scientific theory but to witness the old theories being supplanted. Our grand library is in a perpetual state of revision; there are no sacred tomes, no untouchable truths, and no certainty. There exists only the best description we have of the universe, a description moulded solely by our observations of its wondrous phenomena.

The scientific endeavour, at its core, is inherently humble; it doesn't aspire to uncover universal truths or absolutes. Instead, its essence lies in the pursuit of understanding, and it is within this pursuit that its power and value emerge. Science, without a doubt, has bestowed upon us the modern world, dramatically enhancing our lives. It has not only extended life expectancy and reduced child mortality but has also vanquished numerous diseases and incapacitated many others. Science has granted us the invaluable gift of time, liberating us from the arduous struggle for mere survival and enabling the exploration of the boundless realms of our minds.

In essence, science forms a virtuous circle, with its discoveries generating more time and wealth that, if wielded wisely, can be invested in further voyages of exploration and discovery. Despite its unquestionable utility, I maintain that science is propelled not solely by utilitarian motives but by an innate curiosity. The exploration of the universe and its wonders is as crucial as the quest for new medical treatments, energy sources, or technologies. Ultimately, all these significant advancements rest upon an understanding of the fundamental laws governing everything in nature — from atoms to black holes and everything in between. This is why curiosity-driven science stands out as the most valuable of pursuits, and it is precisely why our journey into the darkness must persist.

1. Conveyors

A fascinating interplay of forces shaping the universe, the role of **Conveyors** appears in the vast cosmic tapestry where celestial activities unfold on unimaginable scales. These cosmic conduits are the silent builders of both cosmic order and chaos, whether they take the shape of electromagnetic fields, gravitational interactions, or other basic forces. They tell the story of the universe by dancing in a way that reflects the cosmic ballet of matter, energy, and information. This chapter delves into the deep impact of these conveyors and examines their role in arranging the cosmos into a magnificent symphony.

✦ THE LEGEND OF LIGHT

Humanity has looked up into the sky throughout recorded history, searching the vastness of the skies for significance. Even if modern astronomy conjures up pictures of advanced telescopes and interplanetary missions, every modern discovery has origins that date back thousands of years to the most basic and basic question of all: what is out there? As it was for our ancestors who reached out to worlds beyond Earth, light is the only means of communication between us and the universe outside of our solar system. Without even the remotest necessity for spacecraft, we can travel from the surface of our home planet to far-off worlds orbiting the Sun by following the path of light.

Gazing upwards is like to peering into the past, as the enduring rays of light that reach us are emissaries from the cosmos' far past. As we enter the twentieth century, we possess the knowledge to unravel the story contained in this ancient light, revealing a tale that reveals the universe's earliest beginnings.

Located near Luxor, opposite the Valley of the Kings on the other side of the Nile, is the Karnak Temple, home to the universal god Amun-Re. Luxor, once known as Thebes, served as Egypt's capital during the affluent and dominant New

Kingdom. With thousands of precisely sized hieroglyphs and an architectural masterpiece from the golden era of ancient Egypt, the 3,500-year-old Karnak Temple is a marvel of engineering and a site of great power and beauty. Its walls could accommodate ten European cathedrals; Notre Dame Cathedral could easily fit inside the Hypostyle Hall alone, an enormous valley of soaring pillars that formerly supported a massive roof.

Throughout human history, religious and ceremonial architecture has served a variety of purposes. While there is clearly a political component—these enormous structures serve to maintain the power of those in charge—it would be a mistake to consider the major accomplishments

of human civilization only in terms of politics. The Karnak Temple is a response to something far older and more majestic. The enormity of the building draws the mind forcibly away from earthly worries and into a realm beyond. Only those with the proper respect for the universe can construct such places.

Karnak is a stone record and a link to the solution to the age-old query, "What is out there?" It is an observatory, a library, and an expression born out of a desire to investigate and cosmological curiosity.

Egyptian religious mythology is a rich and complex tapestry with a large number of temples and tombs, about 1,500 identified deities, and a substantial body of surviving literature. The rich and varied mythology of the great civilization of the Nile is regarded as the most sophisticated religious system ever created. There isn't a single tale or custom from the more than 3,000-year-long dynasty, which reflects the ups and downs of Egyptian culture.

The significance of the waters of the Nile, which serve as this desert civilization's main source of sustenance, is fundamental to both existence and mythology. A verdant stretch of land was gifted along the river by the yearly floods, a striking sight as one approached Luxor from Cairo. But the Aswan Dam broke the long-standing pattern of rising and decreasing waters in 1970, and contemporary irrigation methods keep the lush banks in place. Before the dam was built, summer rainfall in the mountains south of Egypt would cause the Nile to rise and flood until September, at which point the waters would retreat and leave behind fertile soils that support life.

The heart of Egyptian religion was inevitably shaped by the mighty Nile River's significant impact on daily life. Their sky seemed like a vast ocean that the gods travelled in opulent vessels.

According to Egyptian creation legends, there once was an endless primordial ocean from which the earth's single mound formed. The Sun was created when a lotus blossom opened from this mound. Every aspect of this ancient story had a god attached to it. For example, the lotus flower was tied to Nefertem, the god of fragrances, and the fertile soil sprouting from the Nile floods was represented by *Tatenen*, which means **risen land**. The Sun God, who was created from a lotus blossom and has taken on many guises while remaining a central figure in Egyptian religious thought for more than 3,000 years, is the major figure in this faith. All that exists was created by the Sun God, who is the universe's light bearer. The Sun God becomes Amun-Re in the magnificent Karnak, a combination of the local goddess Amun of Thebes and the ancient Sun God Re, rising to unmatched prominence. Egyptian mythology frequently features this inclination to combine gods, which adds to the theological intricacy. Amun is sometimes thought of as the hidden aspect of the Sun, sometimes connected to his nighttime sojourn in the Underworld. Amun is referred to as the "eldest of the gods of the eastern sky" in the Egyptian Book of the Dead, a reference to his rising as the sun god at dawn. Raising to the rank of Amun-Re, he becomes the Gods' King and survives into the Greek and Roman ages as Zeus-Ammon.

The worship of Amun-Re as the ultimate deity became so popular that Egyptian religion nearly reached a quasi-monotheistic status under the New Kingdom. It was believed that Amun-Re-embodied an all-seeing, immortal presence that penetrated all things, surpassing the boundaries of time and space. He resembles the divine conceptions found in Judeo-Christian and Islamic traditions in this way.

The image of Amun-Re that practically covers the walls of Karnak Temple is usually human with a double-plumed crown of feathers; the exact significance of this image is uncertain. He appears in Karnak in animal form, as a ram, but he is most frequently depicted with the Pharaoh.

The temple's alignment with the larger universe is the most breathtaking tribute to Amun-Re at Karnak. The focal point, the **Great Hypostyle Hall**, is positioned with exact precision such that the Sun's disc rises between the imposing pillars on December 21, the shortest day in the Northern Hemisphere, creating a radiant radiance throughout the room. This bright light comes from directly over a little building that was thought to be Amun-Re's residence. Witnessing the solstice dawn while surrounded by the imposing stone columns is a profoundly moving event. It creates a direct link with the historical Egyptian pharaohs, since individuals such as **Amenophis III**, **Tutankhamen**, and **Rameses II** would have stood in the exact same location more over three millennia ago, welcoming the rising December sun.

As the Earth is angled at a distance of 23.5 degrees from the plane of its orbit around the Sun, the dawn position varies every morning. The North Pole in the Northern Hemisphere tilts away from the Sun throughout the winter, keeping the Sun low in the sky. The North Pole gradually tilts toward the Sun as Earth revolves around it, raising the Sun's daily arc across the sky to its zenith around midsummer. Every day, the sunrise position on the eastern horizon shifts as a result of this year-round tilting back and forth.

Observers would notice that, when facing east, the winter solstice marks the location of the most southerly dawn point. After that, the sunrise gradually heads north until the summer solstice, when it reaches its northernmost point. Even while the ancients might not have understood the scientific basis for these phenomena, they would have noticed that the sunrise point stops along the horizon for a few days around the solstices before moving in the opposite direction. For societies who considered the Sun to be a god, the solstices denoted distinct and significant seasons of the year.

Observing the sunrise on this momentous midwinter day from inside the Karnak Temple, the apparent alignment is striking. It is a difficult and contentious endeavour to prove that ancient sites are purposefully linked with cosmic phenomena. Because of its enormous scale and varied variety of buildings pointing in all directions, a temple the size of Karnak will inevitably line up with something in the sky. Two finely carved columns, one on the left and one on the right when facing the rising Sun, flank the structure that houses Amun-Re. This overwhelming evidence has convinced many Egyptologists that Karnak's solstice alignment is deliberate.

These columns are finely carved, and the inscriptions strongly imply a purposeful alignment with the rising sun. The column on the left depicts the Pharaoh hugging Amun-Re, while three carefully carved papyrus stems—a plant found only in the northernmost parts of the Nile—are displayed on one side. On the other side, the identically patterned column on the right shows the Pharaoh hugging Amun-Re while donning the crown of upper Egypt, which is located south of Karnak. The three carved stems on this column are notably lotus blooms, which are a plant that grows only in the south. The deliberate synchronization of the solstice sunrise with the Karnak Temple is strongly supported by this subtle symbolism.

So, it would appear that the columns are arranged and ornamented to indicate the compass directions surrounding the temple, providing strong proof that the structure's central portion is oriented to catch light from a significant astronomical event: the Sun rising in the middle of winter. It is an enormous depiction of the specifics of our planet's orbit around our neighbouring star and its orientation.

The temple is a representation of the ancient Egyptians' obsession with the shifting light patterns in the sky. Although their pre-scientific inclination to worship them was unfounded, the structure also seems to represent a growing understanding of the universe' geometry. An awareness of the Earth's cycles and seasons emerged from studying the sun's changing positions at sunrise. This knowledge was crucial for knowing when to produce and harvest food. Civilizations became richer as a result of the development of more sophisticated agricultural techniques, which in turn allowed them more time for philosophy, mathematics, science, and contemplation. Thus, astronomy initiated a noble cycle in which the pursuit of comprehending the heavens and their significance resulted in intellectual and practical wealth beyond the ancients' comprehension. It took a large portion of recorded human history to go from seeing the regularity in the movement of the heavenly lights to modern science. The work was initiated by the ancient Greeks, but **Johannes Kepler**'s discovery in the seventeenth century provided the accurate explanation of how the Sun, Moon, and planets move across the sky. Although it was challenging to lift the curtain of the divine and see the genuine grandeur of the universe, the benefits of humanity's intrinsic fascination with the stars have proven to be immeasurable.

We have located us during the hundreds of billions of stars that comprise the Milky Way Galaxy by tracking the light. We have measured the chemical compositions of thousands of additional stars in the sky as well as the closest star, Proxima Centauri, during our tour. Even more, we have travelled far into the Milky Way and gazed into the black hole at the centre of our galaxy. However, this is only the start.

✚ WE IN THE UNIVERSE

The vastness of the universe challenges human comprehension and pushes it to its limits. Nevertheless, humans have been able to investigate and comprehend the vast scope of the universe from our modest vantage point on the little planet known as Earth. By means of scientific investigation and technological progress, we have gradually come to understand our place in a cosmic framework extending well beyond the solar system, as we have revealed the great cycles that play out above our heads.

There are an incredible 200 billion stars in our galaxy, the Milky Way, alone. Each star has a unique dance in the cosmic ballet. The magnificence of the cosmos is revealed to us as we continue to study astronomy and explore its mysteries, and it goes well beyond the familiar stars that fill our night sky. The billions of galaxies strewn throughout the cosmic tapestry, each with countless stars and planetary systems, almost make the scale inconceivable.

This cosmic exploration strengthens our bond with a universe that exists beyond our wildest dreams while also deepening our comprehension of the workings of the heavens. From our local solar

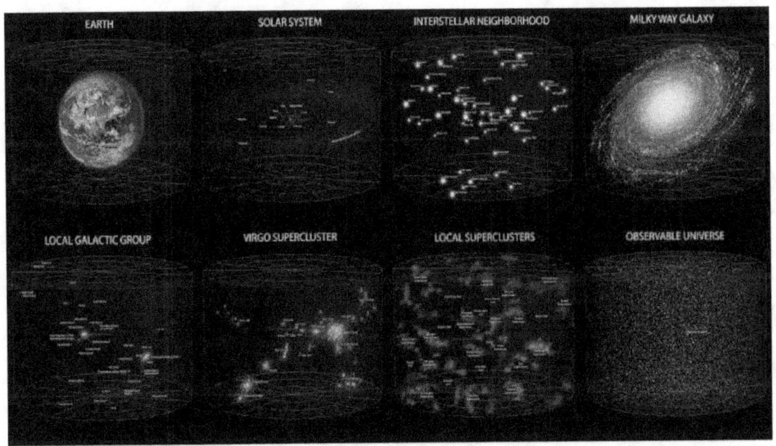

system to the vast cosmic regions, the exploration process serves to further solidify the notion that we are essential parts of a larger framework that extends beyond our immediate celestial neighbourhood. We are invited to investigate, ponder, and be in awe of the wonders that occur on an unfathomable scale by the complex interactions between galaxies, stars, and cosmic occurrences.

✚ OUR NEIGHBOURHOOD

We have a spectacular view to explore our neighbourhood in space from our little earth. The Sun, the star closest to us, is 150 million kilometres (93 million miles) distant, yet hundreds of other stars light up the night sky every night as the Sun fades from view. Up to 10,000 stars can be seen with the naked eye in the most privileged locations on Earth, and they are all a part of the galaxy that is home to us.

A galaxy is an enormous group of stars, gas, and dust that are all tightly connected by gravity. It acts as a theatre upon which the astronomically-scale cosmic life cycles play out. Millions to trillions of stars are thought to reside in each of the estimated 100 billion galaxies that make up the observable universe. The dwarf galaxies, which are the smallest of these galaxies, might have as few as ten million stars, whilst the giants could have as many as 100 trillion stars.

Beyond what can be seen with a telescope, galaxies are thought to contain far more material. They are believed to contain enormous halos of dark matter, a mysterious type of matter that is very different from Earthly matter. Though it is invisible, dark matter has a strong gravitational pull on normal matter and affects how galaxies behave. According to recent estimations, dark matter makes up about 95% of the mass of galaxies, including the Milky Way. This insight kind of pushes the bright stars, planets, gas, and dust to the background.

The enigmatic character of dark matter presents one of the most significant obstacles facing modern physics. The quest to discover the true nature of dark matter promises to be an engrossing adventure as we learn more about the cosmos and add intricacies to our understanding of the cosmic tapestry.

The Greek word **galaxias**, which means **milky circle**, is where the word **galaxy** originates. First used to characterize the recognizable galaxy in our night skies, the Greeks were awestruck by this celestial occurrence even though they were ignorant of its actual size. One of the most magnificent displays of nature used to be seeing the core of our galaxy ascend in the night sky. Sadly, this formerly magnificent nighttime show has been overshadowed by city lights. It now seems to many onlookers like storm clouds breaking the horizon.

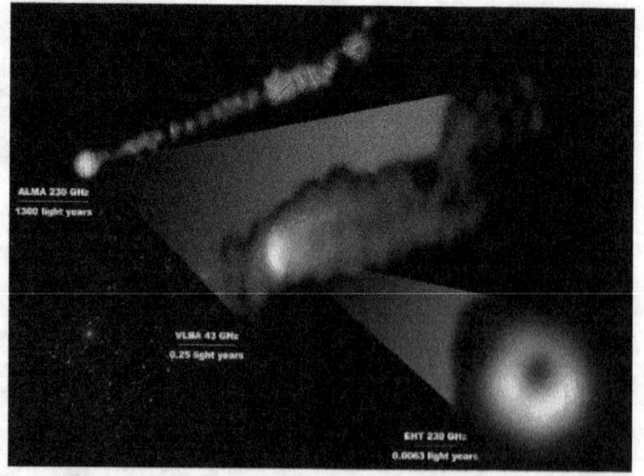

Nevertheless, the first hazy ring of light eventually reveals itself as an exquisite tapestry of stars — billions of them stretching thousands of light years inward — as the Earth rotates each night toward the centre of our galaxy.

Greek mythology explained this ethereal light as Hera, Zeus's wife, spilling milk on her breast to create a thin strip across the night sky. The Milky Way, the name given to our galaxy today, is derived from this ancient story. Interestingly, Geoffrey Chaucer, the medieval poet, is credited with bringing the word into the English language through his writings. Chaucer wrote, "See there, lo, the Galaxyë, Which men clepeth the Milky Wey, For hit is whyt." The essence of the bright streak that gave rise to the current moniker is eloquently described in this poetic description.

✦ MAPPING OUR MILKYWAY GALAXY

Around 200 and 400 billion stars make up the immense cosmic ensemble that is our galaxy, the Milky Way. This estimate is reliant on the discovery of faint dwarf stars, which provide difficulties. Most of these stars are dispersed throughout a disc that has a diameter of around 100,000 light years and an average thickness of about 1,000 light years. It is really difficult to visualize these enormous distances.

To get a sense of the magnitude, consider that a distance of 100,000 light years indicates that light would take 100,000 years to travel across our galaxy at the astounding speed of 300,000 kilometres (186,000 miles) per second. Alternatively, to put the scale into sharper perspective, consider that the distance between the Sun and Neptune, the furthest planet in our solar system, is only about one-sixth of a light day, or around four light hours. It would take almost 220 million solar systems to span the entire galaxy. This enormous space highlights the Milky Way's enormous scale and complexity—the cosmic fabric that houses our neighbourhood of stars.

Scientists propose that there is a supermassive black hole at the centre of our galaxy, and probably at the centre of every galaxy in the universe. Meticulous measurements of the orbit of a star known as S2 lead to this result. **Sagittarius A***, also pronounced **Sagittarius A-star**, is a powerful radio emitter at the galactic centre around which S2 orbits. S2 is the fastest orbiting object known to exist, with an orbital period of slightly over fifteen years and speeds as high as two percent the speed of light.

Astronomers can compute the mass of the object orbiting S2 by accurately estimating its orbital path. Sagittarius A* has a mass that is astounding—4.1 million times greater than our sun. We can see that Sagittarius A* must be closer to S2 than seventeen light

hours, as S2 would collide with it at that distance. A black hole's existence is the only conceivable explanation for how 4.1 million solar masses may be contained in an area less than 17 light hours. Astronomers may declare with confidence that a massive black hole is located at the heart of the Milky Way due to this overwhelming evidence.

More recently, the study of twenty-seven stars—collectively referred to as the S-stars—that all have orbits that bring them near to Sagittarius A* has confirmed and improved these results. These findings broaden our knowledge of the mysterious and potent forces governing the galactic centre.

Beyond the S-stars, the galactic centre appears as a vibrant focus of star activity, with a wide variety of systems that communicate and impact one another. The Arches Cluster, which is thought to be the galaxy's densest star cluster, is one prominent feature. These stars, which are larger than our sun and comprise about 150 young, extremely hot stars, burn brightly but have limited lives—they run out of hydrogen in a matter of million years.

The Quintuplet Cluster, which houses the Pistol Star, one of the brightest stars in our galaxy, is another well-known cluster. The Pistol Star is poised to explode as a supernova (see Chapter 2), nearing the end of its life cycle. In our galaxy, central clusters with the largest star concentration are the Quintuplet and the Arches. Distance from the active galactic centre causes the star density to progressively decrease. Finally, we come across the thin layer of gas in the outermost regions of the Milky Way, known as the Galactic Halo. This outer area highlights the variety and dynamic nature of our cosmic environment by standing in stark contrast to the busy galactic centre.

A star in the Galactic Halo, thought to be the oldest object in the Milky Way, was observed for the first time by astronomers in 2007 thanks to the potent capabilities of the Very Large Telescope (VLT) at the Paranal Observatory in Chile. This old star, designated HE 1523-0901, is a red giant that is nearing the end of its life. The surface temperature of this crimson behemoth is much colder than that of our sun, despite its immense size.

The accurate detection of five radioactive elements within HE 1523-0901—uranium, thorium, europium, osmium, and iridium—makes the star especially fascinating. Astronomers have precisely calculated the age of this old star by using a method similar to carbon dating, which is used by archaeologists to ascertain the age of biological stuff on Earth. When there are several "radioactive clocks" present at the same time, radioactive dating turns out to be quite accurate and trustworthy. An important milestone was the discovery of five radioactive elements in the light released by HE 1523-0901.

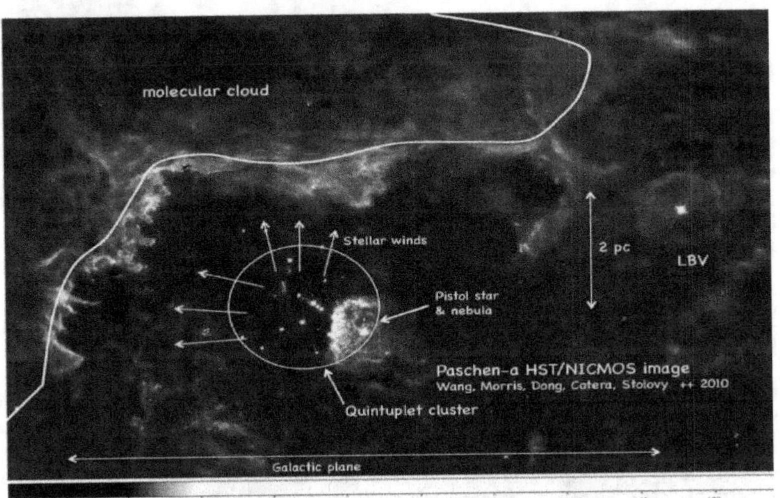

The estimated age of this dying star is an astounding 13.2 billion years, which is nearly as old as the Universe itself, which began little more than 13.8 billion years ago. Since the first generation of stars ended in supernova explosions in the first half-billion years of the Universe's existence, it is thought that the radioactive elements found in this star formed during that catastrophic event (see Chapter 2). This astounding discovery offers a rare window into the far east and new information on the genesis and development of the universe.

✛ THE SHAPE OF THE GALAXY

The Milky Way, our galaxy, is not only big and old, but it also has a fascinating structure. Its structure, which identifies it as a barred spiral galaxy, consists of a disc made up of stars, gas, and dust encircling a bar-shaped centre. From the central bar, this disk gives rise to unique spiral arms that form an enthralling pattern that stretches throughout the galactic abyss.

It was once thought that there were just four major spiral arms in our galaxy: Perseus, Norma, Scutum–Centaurus, and Carina–Sagittarius. Our sun is located in the Orion Spur, a branch of the latter. But new discoveries have shown the existence of a second arm, called the Outer Arm, which is thought to be an extension of the Norma Arm. Our appreciation of the intricacy and beauty inherent in the Milky Way's structural composition is enhanced by this fresh knowledge.

The Sun, the most well-known star in our galaxy, is located close to the inner edge of the Orion Spur. Even though it was originally thought of as an average star, we now know that the Sun is more brilliant than 95% of all the other stars in the Milky Way. The Sun is categorized as a main sequence star because it fuses hydrogen into helium to produce light and energy.

There is an amazing fusion reaction going on in the Sun's core that uses 600 million tonnes of hydrogen per second to produce 596 million tonnes of helium. In the process, four million tonnes of mass that appear to vanish are converted into energy and move closer to the Sun's photosphere. The freed energy is released into the galaxy at this outer layer of the Sun, where it travels throughout the universe as light. The Sun's tremendous energy output and brilliant light greatly add to the cosmic fabric that envelops us.

✚ A STAR BORN

Although our sun is currently in the middle of its life cycle, looking out into the Milky Way's vastness presents a mesmerizing sight: the full cycle of star life is unfolding right before our eyes. Every year or so, a new light appears in our galaxy, indicating that a new star has formed somewhere in the Milky Way. This continuing astronomical event serves as a moving reminder of how dynamic and always changing the cosmic theatre is that is all around us. Every new star that enters our galaxy adds to the cosmic drama that unfolds over the vast expanses of time and space, enriching the tapestry of stellar life.

One example of where stars are born in the universe is the **Lagoon Nebula**. Inside this massive cloud of gas and dust that makes up the interstellar medium, star creation is taking place. This celestial wonder was originally observed by the French astronomer Guillaume Le Gentil in 1747, making it one of the few visible, active star-forming regions in our galaxy.

In the vastness of the Lagoon Nebula, something is slowly happening. Gravity is causing the massive cloud to gradually collapse in on itself. A little bit more matter is accreted into areas with somewhat higher density as this collapse proceeds. These growing clusters eventually attain a critical mass, which starts the transition into bright stars. The Lagoon Nebula, which displays the captivating dance of star birth in the vast Milky Way, is both a celestial masterpiece and a cosmic cradle.

In the centre of this vast astronomical nest, called

the Hourglass, is a mysterious object called Herschel 36. This star's main character is thought to be a **ZAMS** star, which indicates that it has recently begun the main phase of energy production via hydrogen fusion in its core. Recent studies suggest an exciting possibility: Herschel 36 could be three massive newborn stars dancing around one another in orbit, with a combined mass greater than fifty times that of our sun. With this arrangement, Herschel 36 becomes an absolutely massive system of star giants.

Herschel 36 and its stellar companions will eventually come to an end of their cosmic adventure, just like all stars inside the Milky Way will as time unfolds. Many will reach their zenith in a dazzling show of light, ending their lives in a celestial firework of grandeur that adds to the never-ending cosmic cycle of birth and death.

A fascinating celestial show, **Eta Carinae** is made up of large clouds of soaring gas and dust that are remnants of a star explosion that occurred in a star system that is intrinsically unstable. At least

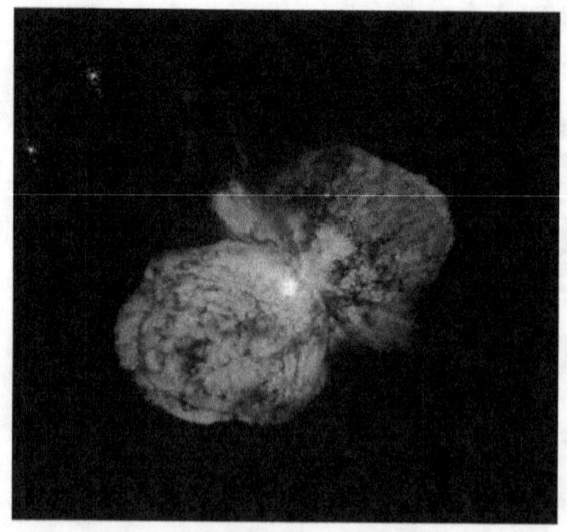

two massive stars make up this cosmic ensemble; taken as a whole, their luminosities are four million times higher than our sun's. Of these stellar components, one is thought to be a Wolf-Rayet star—a monster more than twenty times the mass of our sun that is fighting a losing war to hold on to its outer layers against a fierce solar wind.

Astronomical records from 1843 state that Eta Carinae underwent a catastrophic explosion, making it one of the universe's brightest stars. This star explosion produced a brightness that first confused researchers for a supernova as matter was blasted at velocities close to 2.5 million kilometres per hour. Even though Eta Carinae's fate is predetermined, it miraculously escaped the eruption by hiding deep under the enveloping clouds. The Wolf-Rayet star is fast running out of hydrogen fuel due to its enormous mass, which is speeding it toward final collapse.

It is predicted that Eta Carinae will experience a dramatic change in a few hundred thousand years, which will end in either a supernova or the even more massive hypernova—an explosion of a scale never seen in the known Universe. Eta Carinae is so close to Earth—just 7,500 light years away—that there is a chance that, in the future, its stellar explosion will take the form of a very bright event that can be seen from Earth during the day.

It is astoundingly clear to see the light from these far-off worlds and observe the universe's life cycle in action that light is the ultimate messenger, bringing to us knowledge about the wonders of the cosmos across interstellar and intergalactic distances. Light, however, is capable of far more than just enabling us to glimpse these far-off worlds; it also enables us to travel through time, giving us a genuine and direct link with the past. Not only are the characteristics of light itself able to convey information, but they also make this seemingly impossible condition of things feasible.

In the vast expanse of the Milky Way, the grand spectacle of stellar life unfolds before our eyes. Approximately once a year, a celestial phenomenon captivates our gaze—a new light emerges, marking the birth of a star somewhere within the cosmic tapestry of our galaxy. This cyclical occurrence, a cosmic dance of creation and renewal, perpetuates the ongoing narrative of the Milky Way's stellar inhabitants, as newborn stars join the luminous ensemble that graces the celestial canvas. As each new light ignites, it symbolizes the continuation of the timeless cosmic cycle, where stars are born, evolve, and contribute to the radiant tapestry of our galactic home.

WHAT IS LIGHT?

We set out on a journey to understand the complexities of the world around us, beginning with a basic investigation into the nature of light. Light is the fundamental tool we use to observe Earth, and it is the only way we can travel over the vast reaches of the Universe outside of our galaxy. Even the stars themselves are currently beyond our reach, and our only source of information about their mysterious existence is the light that they emit.

When we go back to the seventeenth century, we enter a critical period when prominent scientists like **Descartes**, **Kepler**, **Hooke**, **Huygens**, and **Newton** focused on studying the characteristics of  light. They were all motivated by the same desire to build better lenses for telescopes and microscopes in order to solve the secrets of the universe on all scales. During this time, scientific and technical developments came together to produce significant breakthroughs in fundamental research as well as profound insights and catalyse one another.

✛ DOUBLE-SLIT EXPERIMENT

Towards the end of the seventeenth century, two opposing hypotheses concerning the nature of light emerged from a deep scientific debate that has shown out to be accurate. In his 1675 essay, the **Hypothesis of Light**, Sir Isaac Newton espoused the idea that light was composed of particles, or **corpuscles**. Newton's scientific rival, Robert Hooke, and the Dutch astronomer and scientist Christiaan Huygens, had opposing views. The majority of physicists supported Newton's particle theory during the ongoing particle-wave controversy that lasted until the early nineteenth century.

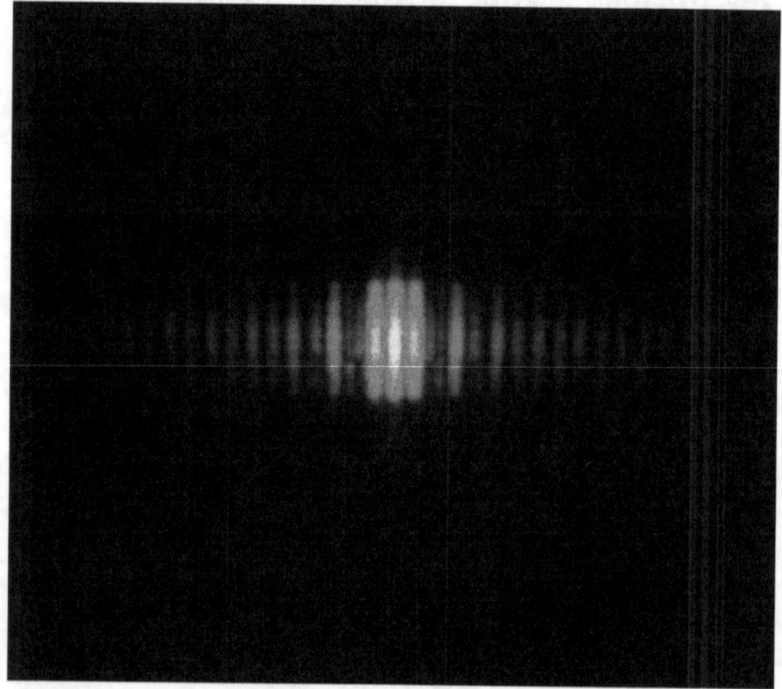

Famous exceptions to the rule, like the renowned mathematician Leonhard Euler, contended that a wave theory was the only explanation for the diffraction phenomenon. The double-slit experiment, carried out by English physician **Thomas Young** in 1801, was a significant discovery that seemed to end the controversy. The experiment provided strong evidence for light's wave nature by proving beyond a reasonable doubt that light diffracted.

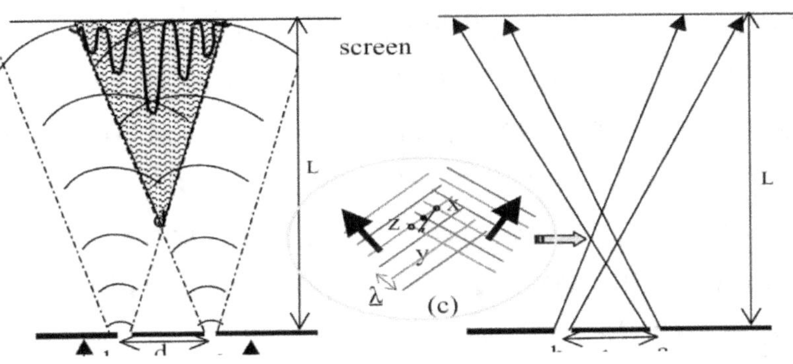

A fascinating and complex phenomenon, diffraction is difficult to explain without referring to waves. Instead of seeing a straightforward bright spot on the other side of the slit when light is channelled through a small slit onto a screen, observers see a complex pattern of alternating light and dark patches. This complex pattern is the consequence of waves interfering with one another; the waves can be strengthened or cancelled depending on how their peaks and troughs line up. The double-slit experiment revealed that light behaved like a wave, which led to more research into the characteristics of these waves.

✦ TEAMS FROM ACROSS THE ENTIRE SPACE OCEAN

In a classic twist of scientific research, the precise explanation of the nature of light came from an unlikely source. Many famous scientists were intrigued by the boom in the study of electricity and magnetism during the mid-1800s. Michael Faraday conducted experiments at the Royal Institution in London, using wires and magnets to manipulate objects intuitively. An electric current is created in a coil of wire when a magnet passes through it, as demonstrated by Faraday's revolutionary discovery. By accident, he created the basis for the generator, which is a vital part of all power plants in use today.

Rather than focusing on useful applications, Faraday's scientific pursuit of understanding electricity and magnetism led him to formulate what is now known as Faraday's Law of Electromagnetic Induction. Simultaneously, André-Marie Ampère, a French mathematician and scientist, made a noteworthy addition by figuring out that there is a force between two parallel wires carrying electric currents. The unit of electric current, the ampere (amp), is defined by this force, which is still in use today. The mathematical description of this phenomena, known as Ampère's Law, has endured over time and influenced many facets of modern life, such as the seemingly insignificant task of replacing a thirteen-amp fuse in an electrical outlet.

Significant progress had been achieved by 1860 in our knowledge of the connection between electricity and magnetism. It was shown that electric currents could be induced by magnets, and that these currents could then affect compass needles in a manner similar to that of magnets. Although a correlation between these occurrences was identified, no thorough explanation has been developed. Thanks to the creative thinking of Scottish scientist James Clerk Maxwell, a breakthrough was made.

In a set of seminal publications that were released in 1861 and 1862, Maxwell developed a comprehensive theory of electricity and magnetism that explained the experimental findings of researchers such as Ampère and Faraday with ease. Maxwell's biggest accomplishment came in 1864 when he published a paper that is recognized as one of the most significant turning points in the history of science.

The work that Maxwell produced in the 1860s was later hailed by Albert Einstein as "the most profound and the most fruitful that physics has experienced since the time of Newton."

Through the integration of electrical and magnetic events into a unified mathematical framework, an astounding prediction was made possible by Maxwell's great insight. This discovery represented a turning point in the history of physics, opening the door to new insights into the nature of the electromagnetic spectrum and light.

Electric and magnetic fields—two ground-breaking ideas—were introduced in order to unite electricity and magnetism. The concept of a field is now fundamental to contemporary physics and is

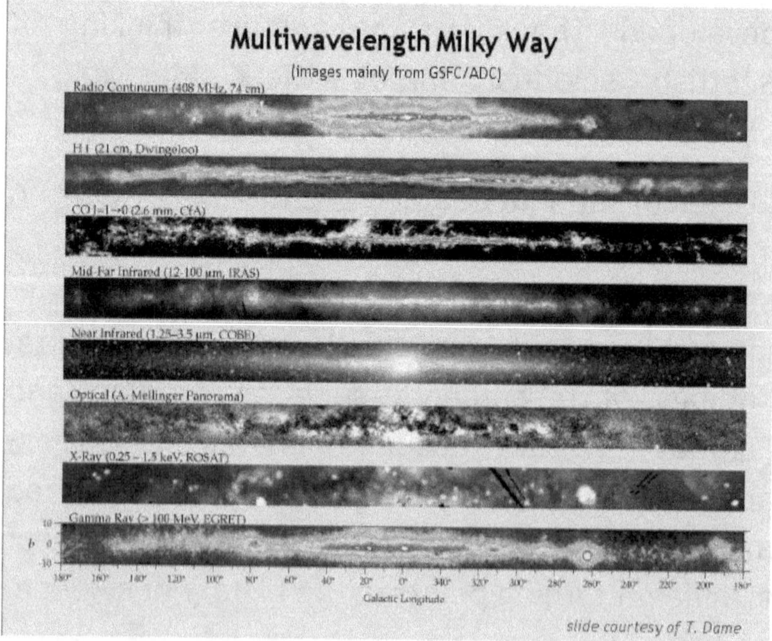

slide courtesy of T. Dame

used in a variety of settings. The temperature in a room is a simple illustration of a field. A temperature field—an array of numbers that represents how the temperature varies across the room—would be created if the temperature were to be measured at various spots and a list was then assembled.

Comparably, the deflection of a compass needle at different points around a wire carrying an electric current is used to introduce the idea of a magnetic field. The magnetic field is the resultant range of values and orientations. Maxwell's genius consisted in realizing that, by stressing electric and magnetic fields, he could create a single set of equations that elegantly explained all known electrical and magnetic processes, despite the fact that this could initially seem abstract and not a simplification. This discovery established the framework for a cohesive comprehension of these essential principles.

THE FOLLOWING EQUATION SUMMARIZES THE RELATIONSHIP BETWEEN ELECTRICITY, MAGNETISM, AND LIGHT SPEED:

A revolutionary discovery was the inclusion of the speed of light, represented by **c**, in Maxwell's equations. The electromagnetic waves, which include light, travel through space at a constant speed equal to the speed of light, according to the formulae. This insight suggested that light is an electromagnetic wave and established a vital connection between electricity, magnetism, and light.

The speed of light, represented by the letter **c** in the equations, could be measured empirically on a lab bench using wires and magnets. It was not only a theoretical quantity. The evidence for light's electromagnetic character was reinforced by this experimental confirmation, which also highlighted the close connections between the many physical processes that Maxwell's equations explain. A significant turning point in the development of physics was the unification of electricity, magnetism, and light under a single theoretical framework.

$$\varepsilon_0 \qquad \text{permittivity of free space } 8.8510^{-12}\,\frac{farad}{m}$$

$$\mu_0 \qquad \text{permeability of free space } 4\pi10^{-7}\,\frac{henry}{m}$$

$$c = \sqrt{\frac{1}{(\varepsilon_0 \mu_0)}} = 2.99792458 \times 10^8\,m/s$$

Maxwell's understanding that his equations might be transformed into wave equations represents a significant insight into the nature of light and electromagnetic. These equations shared the same mathematical structure as those that describe ocean waves or air waves. Maxwell found that the electric and magnetic fields themselves contained waves.

This discovery's main finding is that a shifting magnetic field is produced by a shifting electric field, and vice versa. This self-sustaining wave that can travel across space is created by the dynamic interaction of shifting fields. Crucially, these waves do not require an ongoing source of electric charges to continue oscillating after they are started.

This deep insight gave rise to a cohesive theory explaining light as an electromagnetic wave. It demonstrated the underlying relationship between electricity, magnetism, and light transmission, opening the door to a more profound understanding of the universe's essence. The behaviour of light as an electromagnetic phenomenon was profoundly and elegantly explained by Maxwell's equations and their wave-like solutions.

In addition to demonstrating that light is a wave, Maxwell's equations also predicted that waves move at a particular speed. Maxwell was shocked to see that the speed of these waves matched the speed of light when he estimated their speed. The fundamental conclusion that light is nothing more than oscillating electric and magnetic fields traveling across space was reached as a result of this discovery.

Maxwell's equations state that the ratio between the intensities of the electric and magnetic fields determines the speed of light. This insight established a link between the ostensibly unrelated phenomena of electricity, magnetism, and light and offered a cohesive framework for comprehending their underlying principles. This idea is stunning in its simplicity and elegance, showing how theoretical mathematics combined with practical investigation can reveal the mysteries of the cosmos.

It is important to remember that Maxwell's discovery depended heavily on the measurement of the speed of light. Before Maxwell's time, light transport was understood to occur at a fixed speed as opposed to instantly. This knowledge dates back over 200 years. The first person to measure the speed of light was Ole Romer in 1676, and his discovery laid the foundation for Maxwell's revolutionary theory.

✦ AFTER THE LIGHT'S SPEED

When you open your eyes, the entire world appears before you; light appears to instantly go from object to retina, creating a picture of the surroundings. It is hardly surprising that Aristotle and many other philosophers and scientists thought light travelled *without movement* since light appears to travel at an impossibly high speed. But as the Greek philosophers pondered the nature of light more, a millennium-long argument concerning the speed of light's passage broke out.

The historical context of the speed of light controversy includes prominent individuals like **Galileo, Euclid, Kepler, Descartes, Aristotle**, and **Empedocles**. **Aristotle** held the belief that light travels at an infinite speed, which was endorsed by **Euclid, Kepler**, and **Descartes**. On the other hand, centuries apart, **Empedocles** and **Galileo** contended that light had to have a finite, if very high, velocity.

The logic of **Empedocles** was especially well-crafted. He reasoned that light must exist somewhere in the space between the two places during its voyage if it can travel the enormous distance from the Sun to Earth. This suggests that the speed at which light may move has to be finite. In response, **Aristotle** upheld his theory that light is a presence rather than a moving entity between objects. But in the absence of experimental support, the argument remained unresolved, emphasizing the significance of empirical validation in scientific discourse.

Galileo used two lamps in an experiment to try and determine the speed of light. He held one bulb and placed another at a good distance away for his assistant. Galileo would let the light out of his lamp by opening a shutter. The assistant would open their shutter upon seeing the flash, and Galileo's goal was to time the interval between when he opened his shutter and when he saw the flash via the helper's lamp.

Galileo, meanwhile, was unable to pinpoint the exact speed of light and came to the conclusion that it must travel at an incredibly high speed. He was able to determine a lower bound on the speed of light, though. Because he would have been able to detect a time delay if light had been traveling at a slower pace than sound, Galileo concluded that light must be traveling at least ten times faster than sound. Although he was unable to pinpoint the exact speed of light, the experiment did reveal important insights into how quickly light travels in relation to sound.

In the seventeenth century, Danish astronomer **Ole Romer** conducted the first experiment to prove that the speed of light was not unlimited. Romer was attempting to solve a major scientific and engineering conundrum of the era in 1676: precisely measuring time at sea. Accurate timekeeping was essential for seafarers to traverse the globe safely, but traditional mechanical clocks, which depended on pendulums or springs, were untrustworthy in the choppy seas.

You need both latitude and longitude to find your location on Earth. Romer concentrated on latitude, which is measured in the Northern Hemisphere by measuring the angle of the North Star (Polaris) above the horizon. Determining latitude is more difficult in the Southern Hemisphere since there isn't a star directly overhead the South Pole. Romer used trigonometry and

astronomical observations to try and come up with an accurate method.

While Romer was honing this technique, he noticed something revolutionary. He observed that when Earth shifted in its orbit around the Sun, so did the timing of the eclipses of Jupiter's moon Io. The finite speed of light may be the cause of this difference in eclipse timing. Romer's research was a significant step toward understanding that light is not transferred instantly and travels at a slow speed.

Longitude is much harder to figure out since you need to know the time zone you are in in addition to being able to look at the stars. Zero degrees longitude is the definition of Greenwich in London. When you move west from Greenwich across the Atlantic, your time zone changes, causing it to be earlier in the day in New York than it is in London. On the other hand, your time zone changes as you move eastward from Greenwich, so that it is later in the day in Moscow or Tokyo than it is in London.

On Earth, determining longitude entails figuring out local time, which is typically accomplished by looking at the Sun's location in the sky. The Meridian is the imaginary line drawn in the sky between the north and south points of the horizon that crosses the celestial pole, which is identified in the northern hemisphere by the North Star. At midday, or noon, on its path from sunrise in the east to sunset in the west, the Sun crosses this Meridian, marking the moment when it reaches its zenith in the sky.

Locations separated by fifteen degrees of longitude will experience midday one hour apart because Earth spins once on its axis every twenty-four hours, covering fifteen degrees of longitude every hour. When the Sun reaches the Greenwich Meridian, one would set a clock to read 12 o'clock in order to determine longitude. You are thirty degrees west of Greenwich, for example, if it indicates that it is 2 pm when the Sun is at its peak where you are.

This approach requires accurate timekeeping to function, which led to the creation of precision clocks like marine chronometers for navigational purposes.

✛ SEARCH FOR COSMIC CLOCK

Early in the seventeenth century, King Philip III of Spain offered a prize for the development of an accurate method to determine longitude at sea without using sight of land. Because it was difficult to construct accurate clocks for navigation, scientists looked up to the stars to find high-precision natural clocks.

After discovering Jupiter's moons, Galileo thought he could use their orbits to create a celestial clock. Io, the innermost moon of Jupiter, completes its orbit around the planet every 1.769 days. Jupiter has four brilliant moons that are visible from Earth. Sailors could be able to use Io's daily appearances and disappearances as it passed behind Jupiter as a natural clock for navigation.

Although Galileo's theory made sense, it was difficult to put it into practice, particularly at sea where it was challenging to see Jupiter's moons from a moving ship. Even though he was not awarded the King's Award, his idea established the basis for precise timekeeping on land where stable weather and good telescopes were accessible. An important area of astronomy research was the observation and cataloguing of Jupiter's moon eclipses, especially those of Io. **Giovanni Cassini** gained popularity in the mid-17th century for his research on Jupiter's moons, especially Io. He published tables predicting when these eclipses would be visible from different locations on Earth, and he was the one who first used Io's eclipses to estimate longitude. Cassini dispatched Jean Picard to the **Uraniborg Observatory**, which is close to Copenhagen, where he worked with a young Danish astronomer named **Ole Romer**, to improve these tables.

Romer and **Picard** carefully noted the timings and intervals of more than a hundred eclipses that occurred on Io during the course of several months in 1671. Through this partnership, Romer made the vital

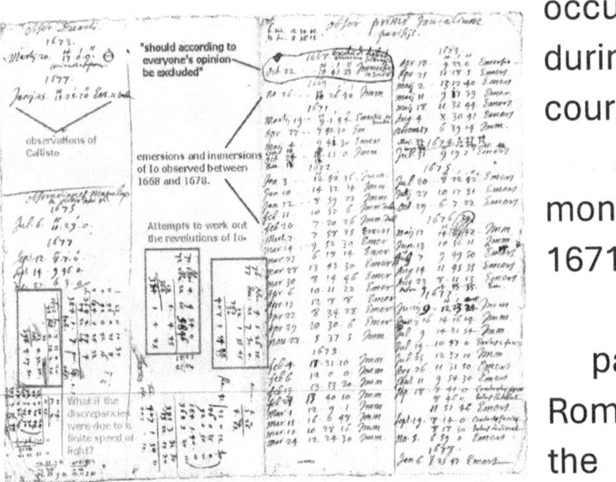

revelation that the moon of Jupiter's accuracy as a celestial clock was not as high as previously believed. Romer discovered a discrepancy in the forecasts for Io's appearance from behind Jupiter when he compared data from Uraniborg with Cassini's views from Paris. There were instances where the difference between the observed and projected timeframes was more than twenty-two minutes. At first, this finding appeared to contradict the idea of using Io as a clock to determine longitude. Romer, however, provided a clever and accurate explanation for the behaviour that was seen.

The apparent variation in Io's eclipse schedule was largely explained by Ole Romer's clever analysis. He came to the conclusion that the temporal difference was caused by Jupiter and Earth's varying orbital distances from the Sun. The inaccuracy relies on the changing Earth-Jupiter distance rather than being inherent to either Io or Jupiter's clockwork.

Romer correctly surmised that the crucial element was light, which travels slowly. Io emerged from Jupiter's shadow later than expected because Jupiter's light took longer to reach Earth as its distance from Earth grew. On the other hand, as the distance shrank, the light arrived at Earth earlier, leading to the unexpected appearance of Io before expected. Through trial and error, Romer was able to account for these varying eclipse timings and determine that light takes about twenty-two minutes to span the circumference of Earth's orbit around the Sun.

Romer did not translate this directly into a speed of light measurement, but his colleague, the Dutch astronomer Christiaan Huygens, estimated the speed of light in his 1678 work *Treatise sur la lumière*. Huygens used an odd unit of measurement to describe the speed: 110 million toises per second. This amount is about equivalent to 299,792,458 meters per second now. This early estimate's main source of error was the incorrect understanding of Earth's orbital diameter around the Sun.

⬇ ROMER'S THEORY

Romer's discovery that the difference in distance between Earth and Jupiter affected the prediction of Io's appearance from behind Jupiter as seen from Earth was a major advance. This knowledge clarified the Jovian clock's apparent oscillations. However, it wasn't until after Romer's passing in 1710 that there was agreement on the speed of light.

Romer's contribution is a two-fold he not only clarified the phenomenon but also provided the first estimate of the speed of light. With this measurement, a fundamental quantity in physics—a constant of nature—was determined. Natural constants, like Planck's constant and Newton's gravitational constant, have been constant since the Big Bang and are essential to understanding the characteristics of our universe. These are fundamental physics constants, and even a small change in their values could result in a very different universe from the one we know today.

⊥ LIMIT OF SPEED

When you talk about breaching the sound barrier, the speed limit you're referring to is related to the sound's speed in a certain medium, like air. The fastest sound wave, or pressure wave, can go through a medium is known as the speed of sound. Sound travels at around 1236 kilometres per hour (768 miles per hour) in air that is 20 degrees Celsius.

Overcoming the speed of sound is the process of breaking the sound barrier, sometimes referred to as supersonic flying. A sonic boom is produced when an item accelerates at a speed greater than the speed of sound. The science underlying this occurrence has to do with how the object compresses air molecules as it passes through the atmosphere, thereby raising the temperature and pressure.

The goal to create aircraft that could go faster than supersonic speeds served as a metaphor for the mid-20th century aviation challenge of breaking the sound barrier. Chuck Yeager's Bell X-1 broke the sound barrier for the first time in aviation history in 1947.

Comprehending the fundamentals of supersonic flight is essential for engineers designing and constructing aircraft capable of reaching such high velocities. Supersonic and hypersonic aircraft design takes into account a number of factors, including aerodynamics, structural integrity, and the capacity to withstand the extreme forces and temperatures involved with supersonic flight.

Imagine you accidentally knock a saucepan lid to the ground. The air molecules beneath it are compressed as it lands, becoming closer together. The air pressure there rises as a result of the congestion.

Consider this rise in air pressure as if it were a tiny, undetectable wave that was emanating from the lid's landing site. This wave resembles the ripples that appear in a pond after a pebble is dropped. The wave or disturbance is what's traveling across the pond, not the water.

Air molecules prefer to equalize pressure, therefore in the case of the lid, they don't really move away from the landing site. Instead, an area of lower pressure is created. Rather, it is this elevated pressure wave or pulse that travels through the atmosphere, like a transient disturbance that is moving away from the site of impact.

We experience sound as a moving disruption of air molecules. Our ears detect pressure changes in the air as sound when anything disturbs the equilibrium, sending pressure waves across the atmosphere. Therefore, the sound you hear when you drop the saucepan lid is really the consequence of the air molecules being moved around in a wave-like manner by the lid.

Numerous factors influence the sound barrier, or the speed of sound in air. It is affected by the mass of the air molecules, which are mostly made up of nitrogen and oxygen, as well as the temperature of the air, which represents the average speed of air molecules. The **adiabatic index**, which measures the air's reaction to compression, also influences the sound wave's speed. Essentially, the average speed of air molecules at a specific temperature determines the speed of sound. Overcoming this fundamental restriction set by the characteristics of the air is necessary to break the sound barrier, often known as achieve supersonic speeds.

The speed of sound, often known as the sound barrier, is the velocity at which a pressure wave passes through the atmosphere rather than a real speed restriction. This knowledge exists before airplanes were created, yet the need to move stuff faster than sound never went away. Throughout World War II, there were several attempts to build supersonic aircraft, leading up to Chuck Yeager's momentous accomplishment on October 14, 1947. Yeager achieved a major aviation first when he broke the sound barrier with the Bell-XS1, being the first person to do so. He broke through the sound barrier and was dropped from a modified B29 bomber to accomplish this accomplishment.

Today, airplanes routinely break the sound barrier, but this routineness frequently hides the complex aerodynamic and engineering obstacles that have to be solved. Test pilot Dave Southwood used the Hawker Hunter, a special aircraft that wasn't intended for level supersonic flight, to demonstrate these difficulties throughout the program's production. The fabled Hawker Hunter was a British jet fighter from the 1950s that could only reach Mach 0.94 in level flight and was not capable of reaching supersonic speeds. But in the hands of experts, it may exceed 1,200 km/h (745 mi/h) and break through the sound barrier. We ascended to 12,800 meters (42,000 feet) and then quickly dropped to the Bristol Channel by doing an inverted dive. The jet breached the sound barrier in a matter of seconds, changing the surrounding air flow and producing a sonic boom that can be heard on the ground.

The mechanics of air molecules and the speed at which pressure waves travel through the air provide the basis for the idea that the sound barrier restricts sound. The matter becomes considerably more complex and fundamentally different when one takes into account the **light barrier**, which is connected to the speed of light. The concept of a light barrier originates from the 1905 special theory of relativity developed by **Albert Einstein**, and it is ingrained in the nature of the cosmos.

Inspired by **James Clerk Maxwell**'s earlier research, Einstein's theory unifies space and time into a single entity called **spacetime**. By proposing that the world contains a fourth dimension—past/future, symbolized by time—in addition to the three well-known spatial dimensions (north/south, east/west, and up/down), Einstein brought about a significant change in our knowledge. The concept of spacetime is based on this four-dimensional structure.

In this spacetime model, the speed of light, or 299,792,458 meters per second (983,571,503 feet per second), is an important quantity. The speed of light is not just a velocity but also a fundamental constant that controls the interaction between space and time, according to Einstein's theory of relativity. It is the speed limit of the cosmos that no mass-containing object can travel faster than. An object's energy and momentum drastically increase as it gets closer to the speed of light, making further acceleration more challenging. Time dilation is a phenomenon that results from time itself being dilated for the moving item.

Essentially, unlike the sound barrier, which is a physical impediment, the **light barrier** is a fundamental limitation originating from the structure of spacetime. The idea that the speed of light is a universal constant has been thoroughly investigated and validated by numerous tests, and it is now regarded as a fundamental concept of contemporary physics.

James Clerk Maxwell's equations for electricity and magnetism and Sir Isaac Newton's classical laws of motion, which had prevailed for two centuries, were incompatible, which made Albert Einstein's revolutionary decision to combine space and time into a single entity, known as spacetime, necessary. When faced with this contradiction, Einstein combined space and time, rejecting the conventional Newtonian understanding of them as separate concepts.

A unique speed that is a part of spacetime's own structure is revealed by Einstein's theory of relativity. Regardless of an observer's velocity with relation to other objects, this speed is a universal constant that stays fixed at precisely 299,792,458 meters (983,571,503 feet) per second. A fundamental tenet of Einstein's theory, its constancy guarantees that all observers will constantly measure the speed of light at this particular value, independent of their relative motion.

This steady speed is essential because it keeps spacetime from becoming aberrant. If the past and future are viewed as equal directions in this universe, why is it so difficult to travel both forward and backward in time? Why is it that advancement is only permitted into the future and not into the past? According to Einstein's theory, spacetime's structure is preserved by the constancy of the speed of light, which serves as a cosmic speed limit.

According to Einstein's theory of relativity, time travel is practically impossible because of the existence of a generally accepted special speed that separates the direction of time from that of space. This unique speed plays a crucial function in forming the structure of our universe and is an essential component of space and time. Its apparent coincidence with the speed of light is surprising. In accordance with Einstein's theory, everything that has mass is restricted to moving through space at velocities slower than this particular value, whereas everything that lacks mass is obligated to go at this particular speed.

Since photons—the light-producing particles—are massless particles, they travel at the speed of light. Since photons are massless and their mass is not well known, there is no compelling reason for light to move at the speed of light. The term **light speed** was only coined to refer to this unique speed that was historically discovered through the measurement of light.

Importantly, the speed of light is a basic attribute of the cosmos that is deeply woven into the structure of space and time, not just an arbitrary restriction. Travel is considered impossible at speeds higher than this one, since mass-containing objects cannot travel at all at these speeds. This intrinsic quality of the cosmos acts as a barrier, keeping the distinction between the past and the future intact and preventing time travel into the past.

✦ TRAVELLING THROUGH THE TIME

In fact, our perception of time is affected by the finite speed of light in a subtle but universal way. Since light travels at a finite speed, everything we see is actually a little bit farther in the past than it appears in real-world situations. For example, you are theoretically viewing a picture from the past when you gaze in the mirror. But this effect is almost imperceptible due to the tiny delay of light over small distances, like the thousand millionth of a second it takes for light to traverse thirty centimetres (twelve inches).

However, the effects of this temporal delay become more evident as we explore the field of astronomy and look up into the expanse of the universe. The delay grows more noticeable at greater distances, and astronomers are faced with the difficulty of seeing objects as they were in the past. This temporal component influences how we perceive the light that celestial bodies send to us, adding a layer of complexity to our comprehension of the cosmos.

When you glance up at the Moon, you are essentially staring at our nearest neighbour from a second ago because it is, on average, 380,000 kilometres (236,120 miles) distant. It is noticeable, but not particularly significant. But when you gaze at the Sun, it seems like you are truly starting to immerse yourself in the past.

At 150 million kilometres, or 93 million miles, the Sun is still very far from Earth even if it is relatively close in cosmic terms. At this distance, the Sun appears closer to us and its finite speed of light becomes more noticeable. It was eight minutes ago. This brings up an interesting phenomenon: the influence of the Sun would last for eight minutes even if it disappeared magically. Its warmth would still caress our faces, its image would still shine in the sky, and we would even keep circling it. Gravitational forces are also subject to the speed of light restriction, which leads to an intriguing interaction between the physical laws controlling the universe and our understanding of it. Therefore, observing the Sun offers us a unique perspective on the dynamics of our solar system and is more than simply a window into its current state. It also takes us on a voyage through time.

The phenomenon of time travel through light gets much more prominent as we look beyond our solar system. Depending on where Earth and Mars are in their orbits, it can take the light from Mars anywhere from four to twenty minutes to get to us. Because of the communication delay, vehicles on the Martian surface, such the Mars Rovers, must move autonomously. This temporal lag has practical ramifications for the design and operation of these vehicles. The communication time to Jupiter, when it is closest to Earth, is approximately thirty-two minutes. When light travels to the farthest reaches of our solar system, it takes around four hours to reach Neptune.

Reaching out into the wide emptiness of space,

Voyager 1, which is presently at the end of the Solar System in interstellar space, takes thirty-one hours, fifty-two minutes, and twenty-two seconds (as of September 2010) to travel a radio transmission round trip. But the time scale changes from hours or days to years when we go outside our solar system to nearby stars. The closest star to Earth that can be seen with the unaided eye, Alpha Centauri, is seen as it was four years ago. The trip through the ages that is made possible by the light from far-off stars gets deeper and longer as we venture farther into the universe.

✚ SEARCHING ANDROMEDA

The intricate science of light and the rich history of the Serengeti are profoundly connected, like a faintly gleaming diamond against the night sky of Tanzania. The vast plains of Africa are bathed in the light of a billion suns since there are no urban lights. A dazzling arc of stars that dominates the sky, the Milky Way Galaxy is made up of too many celestial bodies to count. All of the stunning patches of mist and light spots that are visible to the unaided eye originate from either a star in our own galaxy or the two dwarf galaxies known as the Magallanes clouds, which orbit the Milky Way. But there's one thing wrong with this celestial melody.

Find the characteristic **W** form of the Cassiopeia

constellation, which is located on the other side of Polaris, the North Star, in relation to Ursa Major, which is often referred to as The

Great Bear or The Plough, in order to locate this celestial wonder. Beside Polaris, Cassiopeia is a star that is always visible in the northern skies. It rotates around the pole once every twenty-four hours and never sets at high latitudes. If you see Cassiopeia's **W** standing upright, you will see a quite large, pale, misty area in the sky directly behind the rightmost **V**. It looks duller than the bright stars of Cassiopeia, but being similar in brightness to the surrounding stars.

This seemingly insignificant patch is actually the **Andromeda galaxy**, our closest galactic neighbour, which is the most visually fascinating object to observe with the unaided eye. A trillion suns, or more than twice as many stars as there are in our Milky Way galaxy, are found on Andromeda. It is located approximately twenty-five million million kilometres (fifteen million million miles) from Earth and creates a strong link between the night sky and the far-off universe.

A ray of light set off on an amazing voyage from the Andromeda Galaxy 2.5 million years ago, when our distant relative Homo habilis was roaming the Tanzanian savannah in search of food. This beam of light was moving through the vastness of the

Universe at the speed of light. Generations of prehistoric and modern humans lived across the eons, during which time entire species changed and threatened with extinction. After a while, somewhere along this continuous lineage, one of those ancestors—me—stared up into the sky beneath Cassiopeia, and the light beam ended up on my retina. After two and a half billion years of travel, an electrical impulse was finally created within a nerve Fiber, igniting a series of amazing events in the complex organ known as the human brain. Prior to the light beam's historic journey, this organ did not exist anywhere in the universe.

Unaided eye exploration of the night skies does have its limitations when it comes to solving cosmic riddles. The development of technology has been essential to our ability to go beyond our planet. Sophisticated instruments and telescopes

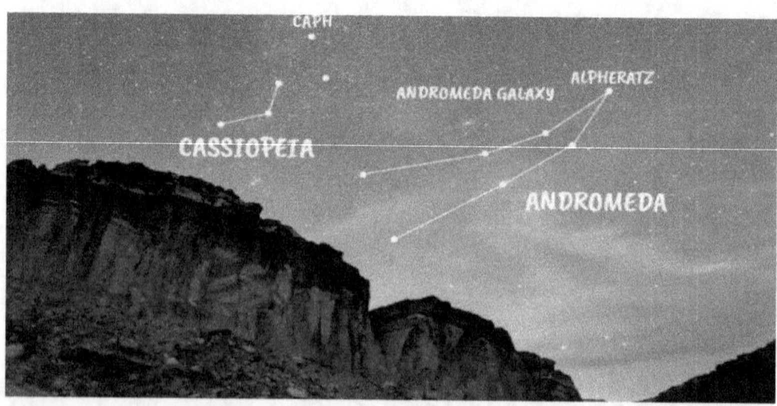

have not only made human space exploration

missions possible, but they have also completely changed our understanding of the universe. With the use of these instruments, we are able to look farther into space and discover more about celestial bodies, far-off galaxies, and the underlying forces that have shaped the universe's vastness. By means of these technological wonders, humanity keeps expanding its knowledge and comprehension while continuing on its path of discovery.

✛ HUBBLE SPACE TELESCOPE

Undoubtedly, our ability to see with our unaided eyes extends only so far into the cosmic past.

When our naked eye could see no farther than Andromeda, our neighbouring galaxy offered us a window into the past that spanned approximately 2.5 million years. But the development of increasingly sophisticated telescopes and astronomy tools has allowed us to see farther into space. With the aid of these formidable instruments, which include ground-based observatories and space telescopes, we are now able to explore much deeper into space and observe light from celestial objects that have travelled over enormous distances over billions of years. By doing this, we are able to access a multitude of data regarding the early stages of the universe and the development of stars, galaxies, and other cosmic phenomena.

An astronomical wonder on par with Galileo's first telescope, the Hubble Space Telescope is a major milestone in the history of astronomy. The Hubble Space Telescope was conceived in the 1970s and named for Edwin Hubble, who made the revolutionary discovery of the expanding cosmos. During its development, the telescope encountered many difficulties.

The project faced obstacles and delays when it was first approved by Congress during President Jimmy Carter's administration, with a projected launch date of 1983. Three years later than anticipated, the eleven-ton telescope was ultimately prepared for launch in 1986. But not only did Hubble's aspirations come to an abrupt end, the entire U.S. space program was brought to a standstill by the terrible occurrence involving the Challenger Space Shuttle's breakup barely seventy-three seconds into its January 1986 flight.

Hubble was then carefully kept for the following four years in a spotless room. During this time, the monthly maintenance costs for the telescope were $6 million, which was a significant amount. The difficulties and setbacks did not lessen the importance of Hubble; on the contrary, they increased the excitement surrounding the spacecraft's ultimate deployment and the revolutionary discoveries it would make in the field of astronomy.

Hubble's launch was eventually accomplished on April 24, 1990, during the STS-31 mission, following the restart of the shuttle program. It was then placed in its intended orbit, 600 kilometres (370 miles) above Earth. Offering unhindered views of the universe devoid of aberrations from Earth's atmosphere was the main goal of the

Hubble Space Telescope. But the first thrill gave way to disappointment rather fast.

After a few weeks of installation, it was clear that Hubble's vision was not as clear as anticipated. When the photos were examined again, they showed a serious optical defect, and more research unearthed a startling fact. The Hubble Space Telescope was launched with a primary mirror that was minutely but dangerously distorted, in spite of decades of planning and a large expenditure. A skewed vision of the universe resulted from the meticulously manufactured mirror, which was supposed to be the most flawless ever built, but it was off by 2.2 thousandths of a millimetre. The telescope's principal mirror imperfection presented a significant challenge and cast doubt on the telescope's capacity to achieve the desired scientific goals.

With the enormous potential and worth of the Hubble Space Telescope, a bold mission was quickly conceived to address its optical problems. As of that time, Hubble was the first and only telescope that was intended to be maintained in orbit by humans. Although it was not possible to install a new mirror, NASA experts came up with an inventive fix. After laboriously calculating the impact of the defective mirror, they concluded that the issue could be resolved by fitting Hubble with corrective optics, which would essentially serve as a set of "spectacles" to improve the telescope's eyesight. This innovative strategy sought to preserve the telescope's purpose and reestablish its capacity to produce precise and undistorted views of the cosmos.

Astronauts on board the Shuttle Endeavour set off on a ten-day mission in December 1993 to install new correction equipment on the Hubble Space Telescope. The amazing, multi-talented astronaut Story Musgrave was in charge of this complex and unusual repair operation. In addition to being a veteran astronaut who had completed four shuttle missions, Musgrave was also a test pilot who racked up an astounding 16,000 flying hours in 160 different types of aircraft. In addition, he was a former US Marine and trauma surgeon with seven graduate degrees.

Musgrave's role in the complex repairs turned into a parable about the space program as a whole. Restoring Hubble's sight, in his own words, symbolized the "majesty and magnificence" of the telescope like a spacecraft or starship. The task entailed giving something that was naturally lovely fresh vitality and restoring it to its intended level of potency. Musgrave stressed that the endeavour encompassed more than just technical fixes; it represented a passion for the Hubble Space Telescope's potential and importance in helping to solve the universe's riddles.

NASA revealed Hubble's revised view to the Universe on January 13, 1994, which was a major turning point in space research. This incident revealed to Earth the astounding beauty of the cosmos in addition as restoring the telescope's capacity to take precise and crisp pictures. Despite the ten-year delay and the hefty $6 billion price tag, Hubble has more than paid for itself by becoming a vital instrument for astronomers and scientists throughout the globe. The updated Hubble Space Telescope is still making ground-breaking discoveries that deepen our knowledge of the cosmos and pique people's interest everywhere.

✦ MOST IMPORTANT HUBBLE DISCOVERY

For almost 20 years, the Hubble Space Telescope has been an innovative instrument for seeing the tiniest celestial lights. It has made it possible for us to rebuild breathtaking sights, giving us a peek into places that are billions of light years away and events that happened billions of years ago—realms that are always out of reach for us physically. One image, the Hubble Ultra Deep Field, stands out among the many that Hubble took because of its significance in illuminating the universe's vastness, complexity, and beauty.

The Hubble Ultra Deep Field took these pictures over an eleven-day period from September 24, 2003, to January 16, 2004, focusing two of its cameras, the Near Infrared Camera and Multi-object Spectrometer (NICMOS) and the Advanced Camera for Surveys (ACS), on a little area of the southern constellation Fornax. It is remarkable that Hubble would have needed fifty of these photographs to cover the entire surface of the Moon, despite the small size of the targeted location. This specific picture has been crucial to our comprehension of the universe, revealing stars, galaxies, and other celestial objects that were previously obscured from view.

The tiny patch of the sky selected for the Hubble Ultra Deep Field may seem completely dark from Earth's surface, but it conceals an amazing mystery. This area was a perfect target for Hubble's observations because there were no visible stars in it. Using its million-second shutter speed, Hubble was able to take pictures of very dim and far-off objects in the shadows. Every minute, a single photon of light strikes Hubble's camera sensors, producing the tiniest objects in the image.

Amazingly, every one of these ostensibly dim spots of light is actually a galaxy—a massive cluster of hundreds of billions of stars. It is possible to see over 10,000 galaxies in the Hubble Ultra Deep Field image alone. If this is extrapolated to the whole observable sky, the observable Universe may consist of more than 100 billion galaxies, each of which has an enormously large number of stars. This finding emphasizes how big and rich the cosmic tapestry is that Hubble has revealed.

The Hubble Ultra Deep Field image is striking even in light of its massive size because of the unusual relationship between the speed of light and the enormous distances between galaxies. The thousands of galaxies that make up this image are all at different distances from Earth, which gives the representation an authentic three-dimensional appearance. But this third dimension is temporal, a representation of time itself, not spatial.

Looking at Hubble's work of art is like peering into a dimension beyond our knowledge, as well as experiencing deep time. Comparable to how an ice core reveals layers upon layers of information about Earth's past, the Hubble Ultra Deep Field offers a chronological tour through the history of the Universe.

The Hubble Ultra Deep Field image presents a breathtaking variety of galaxies with varying ages, sizes, shapes, and colours. Some galaxies are located quite close to us within this cosmic tapestry, while others are located at astronomically large distances. The closest galaxies are around a billion light-years away, appearing larger and more distinct with well-defined spiral and elliptical structures. These galaxies are around twelve billion years old, having originated soon after the Big Bang.

But the real jewels in this picture are the roughly 100 tiny, crimson, asymmetrical galaxies. These galaxies are fascinating not only because of their distinct features but also because they are some of the farthest objects that have ever been seen. More than twelve billion light years separate some of these dim red galaxies. This suggests that nearly the whole 13.8-billion-year history of the Universe has been covered by the light we see from them. Amazingly, the farthest galaxy in the Deep Field—discovered in October 2010—is located more than thirteen billion light years from Earth, so we may observe it as it was just 600,000 years after the Universe initially began.

The photons that make up this old galaxy's picture have travelled a tremendous distance, which is evidence of how big space and time are. There was no Earth or Sun when these light particles set out on their journey from scorching, primordial stars; instead, there was only a chaotic world of dust and nascent stars that would eventually give rise to the Milky Way. Our solar system was created when a whirling cloud of interstellar dust collapsed, around two thirds of the way through its cosmic journey to Hubble's cameras. When sophisticated life first appeared on Earth, these photons were almost there. When the species that built the Hubble telescope appeared, they were almost there.

✛ HUBBLE EXPANSION

Again, we look to the observations of astronomer **Edwin Hubble** to understand why the erratic and chaotic galaxies that he photographed are found billions of light years away. The bright crimson colouring of some of the most distant galaxies that we have seen is a striking element of the image below. I wonder why the majority of these galaxies are red. The solution to this question lies in the contributions of Edwin Hubble.

Using the massive telescope at Mount Wilson Observatory in Pasadena, California, Edwin Hubble studied stars known as Cepheid variables in the 1920s. These stars are a useful tool for astronomers because of their consistent brightness variations over the course of days or months. By watching their fluctuations, astronomers can correctly calculate the true brightness of these objects since the duration of their brightening and dimming is directly connected with their inherent brightness. The primary goal of Hubble's research was to locate and measure the distances of Cepheid variables from Earth in the sky.

During his observations, Hubble came to two important conclusions. First, he proved that the

Cepheid variables, which were previously believed to be bright gas clouds inside the Milky Way, were actually located far outside of it. These variables are found in spiral nebulae. This important discovery proved that galaxies millions of light years away do exist in the universe.

Measuring the light spectrum of stars in spiral nebulae—now known to be other galaxies outside the Milky Way—was Hubble's second significant discovery. Astronomers, including Hubble, discovered that several galaxies showed light that seemed redder than expected during this spectral analysis. Hubble used a measure called redshift to quantify this reddening. When a light's wavelength is longer than anticipated, it is called redshift, and red light has longer wavelengths than blue light.

When Hubble drew a graph connecting the redshift of light from far-off galaxies with their estimated distances based on observations of Cepheid variables, it became clear how significant his second finding was. This graph showed a trend: a galaxy's redshift increases with its distance. Hubble's Law, which explains the connection between a galaxy's redshift, distance, and the universe's expansion, was developed as a result of his observations, which produced strong evidence for the universe's expansion. This law radically altered our conception of the universe.

The unexpected finding made by Hubble was that the graph he created, which connected the redshift of light from far-off galaxies with their separations, roughly followed a straight line. This surprising correlation revealed an astounding interpretation: a galaxy's light gets redshifted, or its wavelengths are stretched, in direct proportion to its distance from us. The universe's fundamental property—its expansion—is the cause of this unexpected correlation.

Space itself is stretching throughout the hundreds of millions of years that light has been traveling over the immensity of the universe to reach us. The distance that light travels and the amount of stretching it experiences are directly proportionate due to the constant rate of expansion. As a result, the galaxies that are farther away have larger redshifts because their light has stretched more significantly over a longer period of time as it travels through the expanding universe.

Hubble's important finding of this **cosmological redshift** marked a turning point in science throughout the 20th century by demonstrating that the universe is expanding and that we live in it. This discovery significantly changed our perception of the universe and established the foundation for contemporary cosmology.

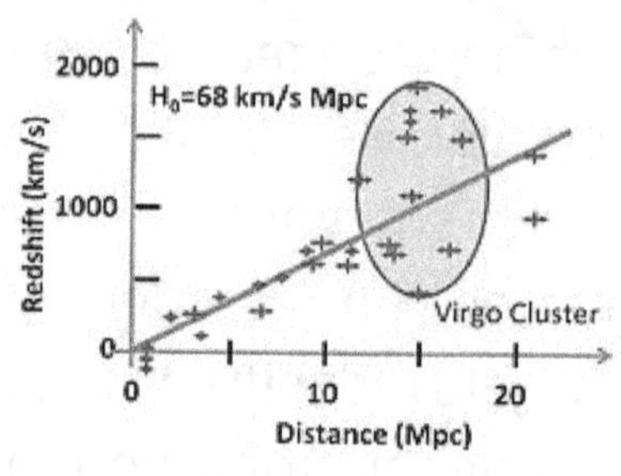

HUBBLE'S LAW: The redshift of the light from far-off galaxies is plotted against their true distance in this diagram, which shows Hubble's Law as a straight line.

CODE EXPLORER

The Free Encyclopedia

✦ THE REDSHIFT

In fact, Edwin Hubble's groundbreaking research at the beginning of the 20th century shed light on the relevance of redshifts in terms of cosmology. He discovered a simple correlation between a galaxy's redshift and distance: the larger the galaxy's redshift, the farther it is from us. The stretching of light as it moves through an expanding cosmos is the cause of this association.

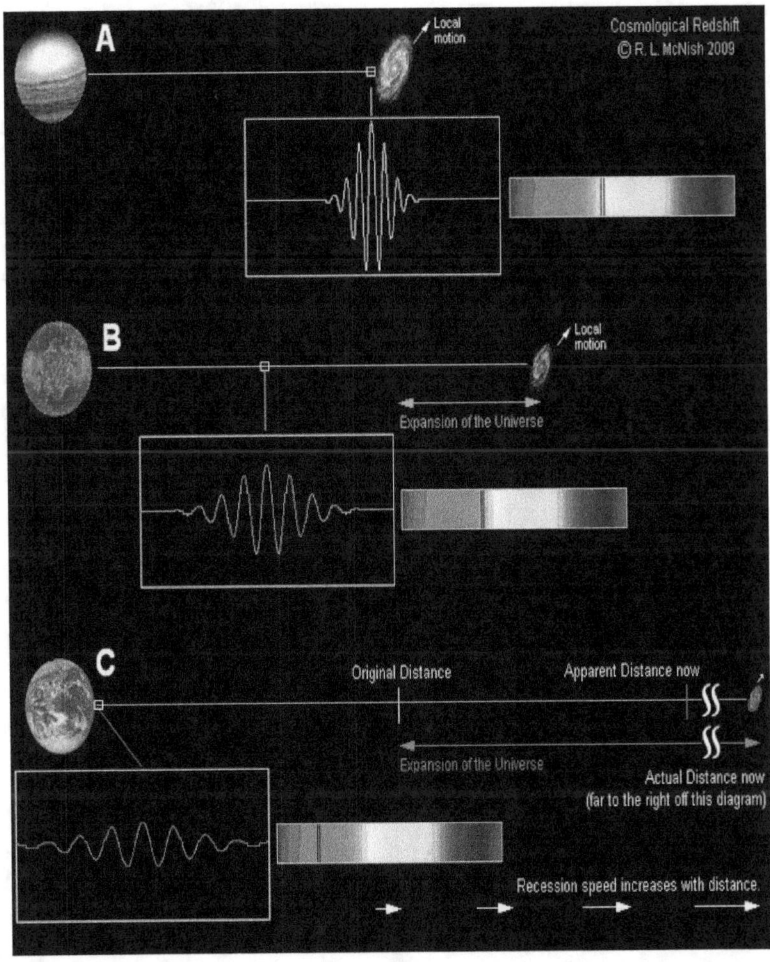

Light travels across enormous cosmic distances, expanding the fundamental fabric of space itself. Redshift increases as a result of the light's stretching over larger distances. Hubble's insightful finding established a connection between the expanding nature of the Universe and the concept of redshifts, so strengthening our understanding of the cosmos and paving the way for future developments in cosmology.

Hubble's redshift-to-galaxies-distance graph offers a plethora of knowledge about the expanding universe. Redshift is defined as the amount of stretching that would be seen if an object were to move away at a specific pace. The Hubble constant is the redshift to distance ratio, which is symbolized by the gradient of Hubble's graph. The Hubble constant is now measured to be around 68 kilometres per second (42 miles per second) per megaparsec. In astronomy, a megaparsec is a unit of measurement that is comparable to 3.3 million light years.

Practically speaking, a galaxy 3.3 million light years away appears to be moving away from us at a speed of roughly 70 kilometres (45 miles) per second, according to Hubble's law. As an example, a galaxy 6.6 million light years away would be receding at a speed of about 140 kilometres per second (90 miles per second), as the velocity grows proportionately with distance.

Furthermore, a value with units of time can be obtained by reversing the Hubble constant. According to a calculation, the Hubble constant is 70 km/s (45 mi/s) per megaparsec. This means that the universe is 14.3 billion years old. It's important to note that, although this closely agrees with the universe's current estimated age of 13.75 billion years, ongoing precise measurements have shown that the universe's expansion does not perfectly follow Hubble's original law. The idea that the universe is expanding faster has been aided by the discovery of dark energy.

One of the most important signs of the universe's expansion is the apparent redshift, which Edwin Hubble and other astronomers saw in the light from far-off galaxies. The redshift suggests that the universe is expanding right now because it indicates that galaxies are moving apart. One can imagine a cosmos whose galaxies were much closer together in the past by projecting this further back in time.

Imagine turning back time and tracking the galaxies as they approach one another, you arrive at a point where the inverse of the Hubble constant indicates that the galaxies would have all converged at that same location. We call this hypothetical event, which occurred about 14 billion years ago, a Big Bang because of its high density and warmth. The deep inference from Hubble's 1920s findings is that there was a massive explosion at the beginning of the universe, which initiated the processes that resulted in the vast universe we see today. Studying the light from distant galaxies and Cepheid variable stars provided all of this information.

It can be difficult to picture the Big Bang, but it's important to clear up the myth that it was an explosion that took place in an already-existing empty space. The conventional wisdom holds that all space, or more precisely, all spacetime, was created at the Big Bang. This suggests that the Big Bang happened simultaneously at every place in the Universe, instead of being limited to a single region.

Thus, when we say that the Big Bang occurred, we mean that it occurred everywhere: in the air surrounding you, in your brain, across the street, in the Solar System, and even in the farthest galaxies.

It happened, in a sense, everywhere in the universe. The universe was always unlimited, if it is infinite now. The important fact is that space was always there at the Big Bang, and that stretching has been a constant feature of space's subsequent growth.

This implies that the galaxies that seem to be moving away from us are not doing so because they were sent into space during the Big Bang explosion. Rather, it is a byproduct of space's continuous expansion, which started with the Big Bang. This viewpoint highlights the intricate and fascinating aspects of cosmology while also challenging our intuitive understanding.

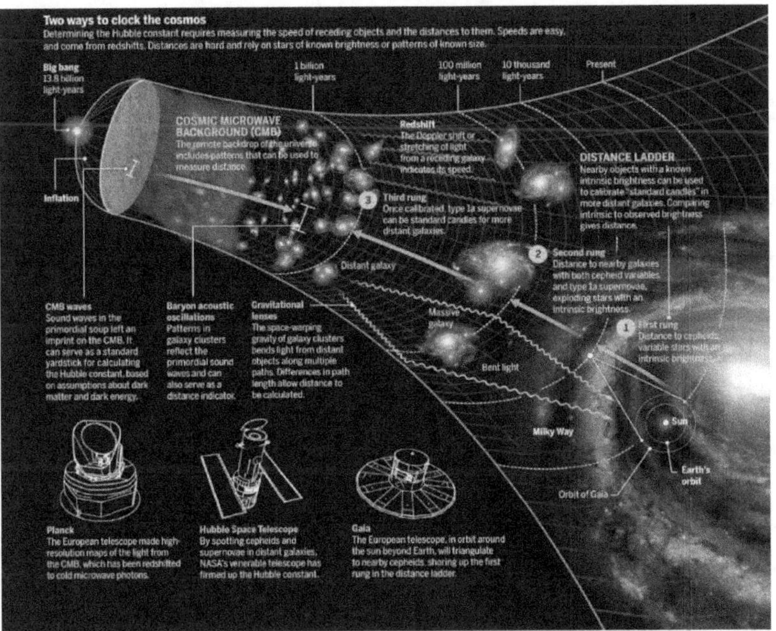

One piece of evidence for the Big Bang is the Hubble expansion; nevertheless, the oldest light in the universe has left us with another, possibly more amazing, imprint of the Universe's catastrophic birth.

✛ BIRTH OF UNIVERSE

The universe is filled with a faint, pervasive glow of radiation known as the Cosmic Microwave Background (CMB). It stands for the leftover heat from the early cosmos, more precisely from the moment the universe become transparent, about 380,000 years after the Big Bang. Prior to this era, photons were scattered by charged particles, resulting in an opaque cosmos that prevented light from moving freely. Photons could move freely because protons and electrons joined to create neutral hydrogen atoms as the universe expanded and cooled. Recombination is the name given to this transition, which is when the cosmos started to show radiation through.

Now, let's explore some key aspects of the Cosmic Microwave Background:

- **Universal Glow:**
  - The CMB is essentially a glow of radiation that fills the entire universe.

- o It is in the microwave part of the electromagnetic spectrum, with a peak wavelength corresponding to a temperature of about 2.7 Kelvin (-270.45 degrees Celsius or -454.81 degrees Fahrenheit).
- **Primordial Light:**
  - o The CMB serves as a snapshot of the early universe, capturing its state when it was only 380,000 years old.
  - o It provides crucial information about the conditions and characteristics of the early cosmos.
- **Variations in Wavelength:**
  - o While the CMB is often described as a near-uniform glow, it actually exhibits minute variations in its wavelength.
  - o These variations, often referred to as anisotropies, are incredibly subtle, representing tiny fluctuations in the temperature of the CMB across the sky.
- **Origin of Anisotropies:**
  - o The origin of these temperature variations lies in the density fluctuations present in the early universe.
  - o Quantum fluctuations during the inflationary period of the universe left imprints in the density of matter, leading

to slight temperature differences in the CMB.

- **Cosmic Archaeology**:
  - Analysing the anisotropies in the CMB allows cosmologists to perform a form of cosmic archaeology.
  - By studying these variations, scientists gain insights into the distribution of matter, the conditions of the early universe, and the seeds that eventually grew into galaxies and galaxy clusters.
- **Evidence for the Big Bang**:
  - The discovery of the CMB and its anisotropies provides compelling evidence for the Big Bang theory.
  - It supports the idea that the universe originated from a hot, dense state and has been expanding ever since.

In summary, a key component of our comprehension of the early cosmos is the Cosmic

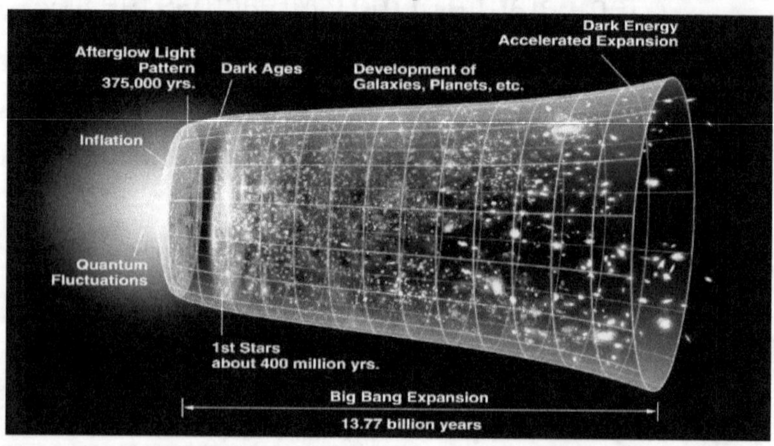

Microwave Background. It is an essential instrument for cosmologists delving into the secrets of our cosmic origins because of its tiny changes, which provide a unique window into the conditions that existed immediately after the Big Bang.

✛ VISIBLE LIGHT

The Namib Desert stretches across southern Africa's west coast. It is the oldest desert on Earth, with a constantly changing sea of sand spanning over 77,700 square kilometres (30,000 square miles) that is altered every minute. The region is an arid wilderness that has defied climate change for more than 50 million years. This is a world moulded by the Sun; its energy propels the wind, which forms gorgeous dunes out of tiny sand grains, and the colours concealed in its light give the surroundings a deep orange hue. The desert is nevertheless alive with light and colour long after the Sun has set, even if it is invisible to the naked sight.

In actuality, a wide range of wavelengths beyond what human eyes can detect as visible light are included in the electromagnetic spectrum. Let's see how the spectrum is expanded to include longer wavelengths, such as radio waves, microwaves, and infrared:

- **Infrared Light**:
  - Beyond the red end of the visible spectrum, we encounter infrared light.
  - Infrared radiation has longer wavelengths than visible light, ranging from about 700 nanometres to 1 millimetre.
  - In everyday experiences, we may feel infrared radiation as heat. Objects emit infrared radiation based on their temperature, and this can be detected with specialized instruments or even felt as warmth.
- **Microwaves**:
  - Transitioning further along the spectrum, we enter the realm of microwaves.
  - Microwaves have wavelengths typically ranging from 1 millimetre to 1 meter.
  - Household microwave ovens use microwaves with a frequency of around 2.45 gigahertz to heat food. The microwave radiation excites water molecules, generating heat.
- **Radio Waves**:
  - Continuing the journey along longer wavelengths, we reach the radio region of the spectrum.

- o Radio waves have wavelengths ranging from about 1 millimetre to kilometres.
  - o These waves are widely used for communication, including radio broadcasting, television, and various forms of wireless communication.
  - o Radio telescopes observe celestial objects by detecting radio waves emitted or reflected by them.
- **Macroscopic Scale**:
  - o As we move further into the radio region, the wavelengths become macroscopic, comparable to the size of mountains.
  - o The ability of electromagnetic waves to have such diverse wavelengths allows for a broad range of applications, from medical imaging using infrared to long-distance communication with radio waves.
- **Common Theme**:
  - o Despite the different names (infrared, microwaves, radio waves), these are all forms of electromagnetic radiation.
  - o The distinction between them lies in their wavelengths, with each portion of the spectrum having unique properties and applications.

From researching astronomical objects in the universe to real-world applications like wireless communication and medical diagnostics, an understanding of the electromagnetic spectrum beyond visible light is essential for a variety of scientific and technical efforts.

Our senses have been restricted to the well-known domain of visible light for a significant portion of human history. But there is a wide range of electromagnetic radiation, including microwaves and radio waves, that extends beyond this limited spectrum. These types of light, which differ from visible light in that they frequently elude human senses, are vital to the transmission of cosmic information. Although radio waves are often associated with technology and communication, they can exist in both manmade and natural forms.

Simple radios can be used as tools for detecting microwaves and radio waves, as they don't require complex equipment. When we tune a radio, we are setting it to receive encoded information carried by light waves rather than to record sound waves. The universe is full with naturally occurring microwaves and radio waves, even though many of the radio waves we come into contact with in our daily lives are intentionally manufactured for communication purposes.

These waves function as messengers from far reaches of the universe, much like visible light does. Microwaves and radio photons transmit precise information throughout the universe, much like photons from the farthest galaxies convey information about their beginnings and travels. With the use of our artificial senses created by technology, we are able to view and comprehend celestial occurrences both close and far away thanks to these naturally occurring waves that offer a unique window into the universe.

About 1 percent of the static, you hear when you tune a radio or hear it between stations is not just noise; it is a signal that is useful to physicists. This static is an echo of the Big Bang, made of stretched light from the very beginning of the cosmos. Originally visible light, the radio waves conveying this information were once a superheated plasma in the early phases of the universe.

Recombination is an important occurrence that happened about 400,000 years after the Big Bang. By that time, the universe had cooled to roughly 3,000 degrees Celsius, which made it possible for electrons to join forces with the nuclei of hydrogen and helium to form atoms. An important turning point was reached when the concentrated plasma gave way to a state where atoms might form. At last, light could travel across space without interference, and the cosmos, a mere tenth of its current size, radiated visible light at a temperature comparable to that of red giant stars' surfaces.

The old light from this era has had the chance to travel through space while the cosmos has continued to cool and expand. But the light has stretched significantly because of the universe's expansion, passing through the visible and even infrared regions of the spectrum. We can now detect this ancient light in the electromagnetic spectrum's microwave and radio portions. The **Cosmic Microwave Background** (CMB) is a long-wavelength, weak glow that was detected in 1964 by **Robert Wilson** and **Arno Penzias**. A significant portion of the supporting data for the Big Bang theory came from the CMB.

✚ CAPTURING THE PAST

Launched on June 30, 2001, from the Kennedy Space Centre in Florida, the Wilkinson Microwave Anisotropy Probe (WMAP) was a specialized telescope whose sole purpose was to detect the faint glow of the Cosmic Microwave Background (CMB) and produce the first-ever image of the cosmos. WMAP yielded important data over nine years, providing a thorough look into the early cosmos and its later expansion and evolution. Notwithstanding its recent retirement, scientists are still devoting a great deal of time and effort to studying the rich details that WMAP's photo offers about the early cosmos.

While the Wilkinson Microwave Anisotropy Probe (WMAP) image may not have the visual appeal of a spiral galaxy or nebula, astronomers consider it to be among the most important images of the sky ever obtained. Its tremendous volume of knowledge concerning the history of our cosmos accounts for its significance. Beyond its outward look, this picture is an invaluable resource for solving cosmic puzzles and expanding our knowledge of the universe's evolution.

The first raw image from the Wilkinson Microwave Anisotropy Probe (WMAP) shows our Milky Way

Galaxy glowing brightly, creating a noticeable ring across the sky. Nevertheless, painstaking efforts to weed out irrelevant information—including observational side-effects—lead to a simpler but incredibly significant and educational image. This processed picture, which is seen below, acts as a snapshot of the night sky and provides an unprecedented amount of information on the composition of our universe during the recombination event.

Continuous modifications to this image have been made over the nine years of WMAP's operational lifetime. With each refining, more complex details hidden in the old light have come to light, advancing our knowledge of the origins and development of the universe.

An image of the sky's temperature is created using the WMAP data. Any given point's observed light's wavelength can be used to determine its temperature; shorter wavelengths indicate greater temperatures, while longer wavelengths indicate lower temperatures. Just 0.0002 degrees separates the red and blue regions in terms of temperature. The CMB has an average temperature that is 2.725 degrees Celsius above zero. That is equivalent to 2.725 K, or -270.425 C, on the Kelvin temperature scale.

The Cosmic Microwave Background (CMB) image has minute temperature variances that are extremely significant because they indicate that in the early cosmos, certain parts of space were somewhat denser than others. These variations are important even though they are subtle; they represent the original seeds of galaxies. The red patches in the CMB represent regions of the universe that at the time of recombination were, on average, around 0.5 percent denser than their surrounds.

Because of their higher density, these denser regions grew slightly more slowly than their less dense surrounds when the universe expanded as a whole. In essence, as a result of their slower expansion due to higher density, their density rose even more in relation to the surrounding space. This was made possible by their increased gravitational impact. These regions would have been twice as dense as their surroundings by the time the universe was just over a billion years after the Big Bang, or one-fifth of its current size.

By this time, the stuff in these denser areas cooled down enough and became dense enough to start a gravitational collapse. The first stars formed and the cores of galaxies, including our own Milky Way, emerged as a result of this collapse. This cosmic era is the time span that is represented in the most redshifted data from the Hubble Space Telescope, which shows how the first galaxies formed. These seeds eventually bloomed into the huge cosmic structures we witness today; the telltale evidence of these seeds are the minute oscillations seen in the CMB.

Then the universe began to light up as many stars burst into flame, casting their light throughout the cosmos. Generations of stars lived and eventually died over billions of years. A star that would become known as the Sun was born in a little area of space known as the Orion Spur off the Perseus Arm inside the Milky Way galaxy. This completes the story of how the solar system originated and ultimately returned to the dense areas of space that first appeared in the early universe.

Still unanswered is this: What is the source of the tiny density variations we see in the Cosmic Microwave Background (CMB)?

The modern inflationary model of the very early Universe is possibly the most fascinating idea in science. The universe had an incredible phase of expansion around 10^(-36) seconds after the Big Bang, during which its volume increased by an unfathomable factor of roughly 10^78. Said another way, this translated into a volume rise of a million billion billion billion billion billion billion billion billion billion within a million million million million million millionths of a second following the Big Bang. This quick expansion came to an end after about 10^(-32) seconds.

The observable portion of the universe, which makes up our night sky's hundreds of billions of galaxies, would have been far smaller before inflation than a single subatomic particle. Quantum physics rules at such small distance scales, and the denser regions seen in the Cosmic Microwave Background spectrum would have formed from minute quantum fluctuations before to inflation that were amplified by the fast expansion.

The CMB provides a window into a period of the Universe's life that is significantly earlier than 400,000 years after the Big Bang, if the inflationary theory is correct. It enables us to see the traces of what happened in the unbelievable first million million million million million millionths of a second following the beginning of the universe. The great power of science is demonstrated by the astounding notion that humans, living on the surface of a rock at the edge of a galaxy, can comprehend the evolution of the Universe and rationally conjecture about the origin of time itself. This capacity to decipher signals sent by light beams across the universe reveals an unmatched rich cosmos.

This narrative has one more twist. Light has served as our journey's messenger, bringing tales of far-off lands and the distant past to our shores. However, evidence from one of the planet's early locations suggests that light may have been much more than just a muse throughout human history.

✛ FIRST SIGHT

Nestled in the Canadian province of British Columbia, the Burgess Shale is one of the planet's most important and poignant scientific sites. This area was under the surface of a primordial ocean around 505 million years ago, when a major mudflow occurred. Everything in its path was buried by this mudflow, providing a unique image of an amazing juncture in Earth's evolutionary history.

The mudflow functioned as a time capsule, remarkably precisely maintaining an entire ancient ecosystem. This lucky geological event preserved the life of these primitive creatures from a bygone era, much as the Egyptians painstakingly created their magnificent tombs half a billion years later. This ancient treasure trove was hidden for hundreds of millions of years until it was found in 1909 on a mountaintop.

This geological wonder, known by the name Burgess Shale, provides insight into the complexities of life at a critical juncture in Earth's history. Our understanding of the early life forms' diversity and evolution, as well as the mechanisms that impacted life on Earth in the past, has been greatly enhanced by the fossils found in the shale.

The Burgess Shale is one of the most important fossil sites in the world because of the incredible age of the animals as well as their vast quantity and diversity. Fossils of sophisticated living forms are absent from the surface of the Earth until about 540 million years ago. Even if there was life in this earlier epoch, it was basic, skeleton less species that are not represented in the fossil record.

The Cambrian Explosion, also known as the Evolutionary Big Bang, was a spectacular event that happened in the geological blink of an eye during the Cambrian Era, as recorded in the Burgess Shale. During this time, a wide variety of sophisticated multicellular life on Earth began to appear. The fossils from the Burgess Shale offer a rare and priceless window into this revolutionary time, enabling researchers to examine and comprehend the wide range of living forms that arose during the Cambrian Explosion. The Burgess Shale fossils have a major impact on our understanding of the swift evolutionary changes that moulded the planet's biological environment.

That being said, what led to the evolution of complex life? These fossil beds, located high in the Canadian Rocky Mountains, hold a hint. The image on the left depicts a trilobite, a sophisticated organism that is one of the ancient species that can be found here. Despite having jointed limbs and external skeletons, trilobites—which are now extinct—were most notable for having intricate, compound eyes. These ancestors' keen vision allowed them to track their prey with great efficiency, as well as recognize forms and movements. These trilobites were incredibly successful creatures because of their capacity to sight; they lived for a quarter of a billion years before going extinct 250 million years ago during the Permian mass extinction.

According to a widely accepted view, the evolution of eyes in animals like trilobites was crucial in the Cambrian Explosion, often known as the Evolutionary Big Bang. Natural selection takes on a new dynamic when eyes appear in a predator. Due to their considerable hunting advantage, ocular predators put pressure on their victims to devise survival countermeasures.

The pressure of natural selection drives the development and improvement of eyes in different species, sparking the start of the evolutionary arms race. Predators continue to evolve increasingly complex vision, which sets off a vicious cycle in which prey creatures improve their defences. Due to the interaction between predators and prey, which is facilitated by the development of eyes, living forms experience a feedback loop that gets more complicated.

Essentially, an organism's ability to see initiates a series of evolutionary adaptations in response to the selective pressures imposed by this newfound ability. This is how eyes evolved in one species. It is believed that this dynamic process played a major role in the rapid diversification of life during the Cambrian Explosion.

Among the first living things to engage with and make use of the light that pervaded the cosmos were the ancient organisms preserved in the Burgess Shale. The Sun and stars performed a cosmic ballet in the night sky that no one noticed before they were born. These extinct species are thought to be our ancestors. Interestingly, there is evidence in the Burgess Shale that points to the possible critical role that a strange worm-like creature known as **Pikaia** had in the evolution of modern vertebrates, including humans.

Though it doesn't look very remarkable, Pikaia is important to evolutionary biology. Pikaia may be the earliest known ancestor of contemporary vertebrates, according to some scientists, placing it on our evolutionary tree. The idea that Pikaia may have had light-sensitive cells further heightens the curiosity. By enabling the organism to perceive light, these cells might have given it a survival advantage and helped it avoid predators in the Cambrian seas.

This suggests an intriguing theoretical relationship: if Pikaia's light-sensitive cells were real, they may have evolved over hundreds of millions of years to give rise to eyes in more sophisticated animals. There is a theory that suggests this rudimentary capacity for light detection may have contributed to the eventual appearance of humans on Earth. Our ability to look up and think about the world is a sign of our ability to sense and interpret light, which has been essential to our comprehension of the cosmos.

It is true that comprehending the universe is like solving a mystery, and light has carried the key pieces of information to us over great distances in both space and time. We have been able to witness cosmic phenomena that our ancestors would have been unable to comprehend, such as the emergence of stars in remote corners of the cosmos, galaxies existing at the outermost limits of the visible universe, and glimpses of our cosmic origins taken moments after the universe began. All thanks to our ability to capture and interpret this light.

The idea that our uniquely human characteristics are the result of an evolutionary journey from the primitive biological light detectors that first appeared on Earth approximately half a billion years ago, during the Cambrian Explosion, is an intriguing one. After evolving over eons, our green, blue, and brown eyes can now look up into the night sky, catching the light from far-off stars and telling the story of the cosmos. It is evidence of the close relationship between the evolution of life on Earth and our efforts to understand the enormous universe.

2. Galactic Dust

In **Galactic Dust**, Chapter 2, the perennial query of what the universe is made of is addressed. The question becomes more profound in the midst of the remains of the Big Bang, the first cosmic explosion that produced a blazing hot, highly structured fireball devoid of any distinguishable form. How did the basic components of the human being evolve from this primordial chaos? We are going to take a fascinating trip to solve the cosmic riddles that lay at the heart of our existence as we explore the complexities of Galactic Dust.

✚ ORIGINS OF HUMANITY

What components make up humans? Throughout ancient times, philosophers and scientists have been devoting a great deal of effort to finding a solution to this topic, which may be among the oldest. The hunt for the universe's building blocks is still ongoing, so by the time you read this book, there might have been another chapter added to the tale. That is the strength, thrill, and speed at which contemporary science is developing. This chapter tells the tale of how those building pieces were formed in the very beginning of the universe, fused together into increasingly complex structures in the space furnaces over billions of years, and then painstakingly put together by the forces of nature to become planets, mountains, rivers, and people.

In 400 BC, the ancient Greeks—Leucippus and Democritus in particular—delved into the profound topic of what the universe's basic building blocks are. Despite the absence of current scientific methods and technological developments, their contemplations resulted in the development of the atomic hypothesis. This idea held that the world was made up of an endless variety of indivisible and indestructible atoms, each with a distinct shape that allowed them to fit together delicately to form the different substances that make up the physical world.

According to their understanding, distinct atom types were associated with distinct materials; iron, for example, was made up of one sort of atom, water, another, and human flesh, a third. These atoms were said to have properties that mirrored those of the substances they made up of; for instance, water atoms were thought to be slick, whereas metal atoms were thought to have shapes that allowed them to interlock, giving metallic substances their hardness.

Though it admitted that the world was made up of smaller components, the antiquated atomic hypothesis was thought to be unduly complicated and, in the end, incorrect. It is now clear from modern science—especially particle physics— that an endless variety of atom types is not necessary to account for the complexity of the world we live in. Rather, the fundamental constituents of rocks, fish, Earth, and the sky are all the same for people.

Particle physics studies these building blocks and how they interact, and modern experiments, like those carried out at the Large Hadron Collider at

CERN in Geneva, aid in our continuing effort to list and understand the elements that make up the universe.

By early 2011, humankind had evolved to distinguish twelve elementary particles as the fundamental building blocks of the universe. Surprisingly, only three of these particles are needed to build all that exists on Earth, including human bodies. The electron, the down quark, and the up quark are these three fundamental particles.

- **Up Quark**: The up quark is one of the six types of quarks, which are elementary particles that combine to form protons and neutrons. A proton, for example, is composed of two up quarks and one down quark.
- **Down Quark**: Similar to the up quark, the down quark is another type of quark. It combines with up quarks to create protons and neutrons, the building blocks of atomic nuclei.
- **Electron**: The electron is a fundamental particle that orbits the atomic nucleus. It carries a negative electric charge and plays a crucial role in the structure of atoms.

These three elementary particles can be assembled into more familiar particles such as protons and neutrons. For instance:

- Two up quarks and one down quark combine to form a proton.
- Two down quarks and one up quark combine to form a neutron.

The basis for the wide variety of chemical elements present in the universe is made up of protons, neutrons, and electrons in turn. The atomic nuclei of the ninety-four naturally occurring chemical elements differ in the number of protons they contain. These fundamental chemical elements include iron, gold, silver, hydrogen, carbon, and oxygen.

✝ THE LIFE CYCYLE

One of the world's holiest rivers begins fifteen miles northeast of Kathmandu, the capital of Nepal, when three minor streams converge. The Bagmati is a swift mountain stream at its source, but it transforms into a broad, magnificent river as it passes through the Kathmandu valley and approaches the famous Himalayan city.

One of the holiest places in Hinduism, the fifth-century Pashupatinath Temple, is located in the eastern section of the city, at the epicentre of the river's legendary power. Travelers from all over India and Nepal travel to these places to worship and honour the god, Shiva.

The holy landmark of Pashupatinath, which is tucked away in the centre of the colourful and varied Hindu faith, has long enthralled my senses. Hinduism, which is deeply rooted in both philosophy and mythology, develops like a convoluted story in which temples, sacred locations, rituals, and the fabric of daily existence all come together to create a bewilderingly beautiful mosaic.

Pashupatinath's soul becomes apparent to me as I stand among its kaleidoscope of colours—a joyous assault on the senses. There's a noticeable blend of vivid colours and ethereal undertones in the air. The vibrant hues and sounds of India blend in perfect harmony with the calm philosophy that comes from the highest altitudes of Tibet. The lines separating the two worlds merge here, blending into the glistening atmosphere of the towering Himalayas.

The holy and the mundane merge harmoniously in Pashupatinath. There is something about rituals and spiritual dedication that is in the thin, crisp air. The tinkling of bells and the rhythmic clang of market vendors meld together in the complex soundtrack. The pleasant aroma of incense blends with the smell of burning flesh, a sombre reminder of life's transience, to create a sensory symphony that lingers in the mountain air.

Pashupatinath's spiritual fabric transcends the material world and touches legend. The chanting of the Monkey Gods, protectors of this hallowed place, breaks the sombre calm. The sound of their chants reverberates against the background of antiquated ceremonies, each one serving as a monument to the eternal union of the divine and the material world.

However, despite the colourful mayhem and serious ceremonies, Pashupatinath exudes a deep tranquillity. As the smoke from a thousand funeral pyres rises, it carries souls and prayers up to the sky. Life and death dance together in this hallowed place, and the essence of Hindu philosophy unfolds in the soft Himalayan breeze, beckoning reflection at the nexus of the mortal and the holy.

At Pashupatinath, the holy river and the historic temple serve as physical representations of the

Trimurti, the divine trio that embodies the essential elements of the Supreme Being. This is where the deep philosophy of Hinduism is revealed. The three deities representing creation, preservation, and annihilation in the cosmos are Brahma, Vishnu, and Shiva.

As the cosmic creator who shapes the fundamental fabric of the Universe, Lord Brahma is depicted in the holy tapestry of Hindu mythology. The preserver, Lord Vishnu, ensures that life continues by guarding the delicate balance of existence. But Lord Shiva, who stands for evil and the end of the world as we know it, is the one with the powerful ability to destroy.

However, Shiva's destructive power is not seen as an inevitable end in the teachings of Hinduism. Rather, it is acknowledged as a fundamental component of the never-ending circle of existence. The old must be destroyed in order to make room for the new, in a cosmic dance of rebirth and destruction. In this framework, Shiva appears as a regenerating force as well as a herald of darkness, continuing the never-ending cycle of death and rebirth that lies at the heart of Hindu mythology.

Standing guard by the river, the Pashupatinath Temple transforms into a hallowed place where death and life meet. The temple is seen as a portal to transcendence, and the soft flowing of the river echoes the beats of the cosmic dance here. This holy site takes on a special significance because of the belief in the cyclical nature of existence, which holds that death is not an end but rather a transitory stage in the never-ending cycle of cosmic regeneration.

The devotees enjoy the profound wisdom embodied in the Trimurti's elaborate dance, which spans creation, preservation, and the essential destruction that makes room for regeneration, as they assemble along the holy river and within the temple's sanctum. Pashupatinath, representing the everlasting ebb and flow within the complex web of Hindu thought, is a monument to the peaceful coexistence of life and death because of its strong link to the essence of Lord Shiva.

Hindus hold that the goal of a soul's existence on Earth is to perfect itself via a cycle of rebirth and reincarnation. It can only then be released from its corporeal existence and reunited with the Universal Soul. According to the Bhagavad Gita, "the soul discards worn-out bodies and wears new ones, just as a man discards old clothes and puts on new ones." It is thought that your soul would be freed from the exhausted body as soon as possible by having your body burnt on the riverside next to Shiva's Pashupatinath Temple.

The deceased must be immersed in the Bagmati River three times before to cremation, per Nepalese Hindu . customs. Following the cremation, the principal mourner, who is typically the deceased's first son, is required to take a quick wash in the holy river and light the funeral pyre. Numerous family members who participate in the funeral procession also take a dip in the river or sprinkling holy water on themselves. Because of this, the riverbank is an odd and congested location. While it may seem frightening to my British eyes—death is rarely, if ever, exhibited like this—walking between the funeral pyres while friends and family go through their customs is not viewed as inappropriate in Kathmandu.

The five components of the human body are air, water, fire, earth, and ether, according to Hindu religion. Interestingly, their view about what happens to these elements after death is similar to our current understanding of how the world functions, even though modern science finds this to be overly convoluted.

The belief that the body's components are returned to Earth to be recycled and repurposed is the foundation of the cremation ritual. Therefore, death is only the end of one stage of existence and the beginning of another; it is a part of a natural cycle of death and rebirth, not the end for the immortal soul or the mortal flesh. Regarding the atoms and chemicals found in human bodies, contemporary science fully concurs with such notion. My components won't be miraculously eliminated when I pass away; instead, they will be brought back to Earth and, given enough time, will combine to form another structure.

Hinduism is not the only religion, of course, that has elaborated and poetic tales concerning the beginnings and development of Man and the Universe. A creation narrative explaining our origins, how we got to be here, and what happens to us after we die is at the core of almost every civilization and religion on the planet. This implies that our natural tendency to be curious about our beginnings may perhaps be a characteristic of what makes us human.

Like the great philosophical traditions of antiquity, modern science has a physics and cosmology-based genesis myth of its own. It can tell us the composition of the planet and our origins; in fact, it can teach us the composition and origins of everything in the world. Because telling this story requires understanding the history of the Universe, it also satisfies one of the most fundamental human needs: the yearning to be a part of something far bigger. It also teaches us that knowing the lives and deaths of the stars is a more direct route to enlightenment than knowing our own.

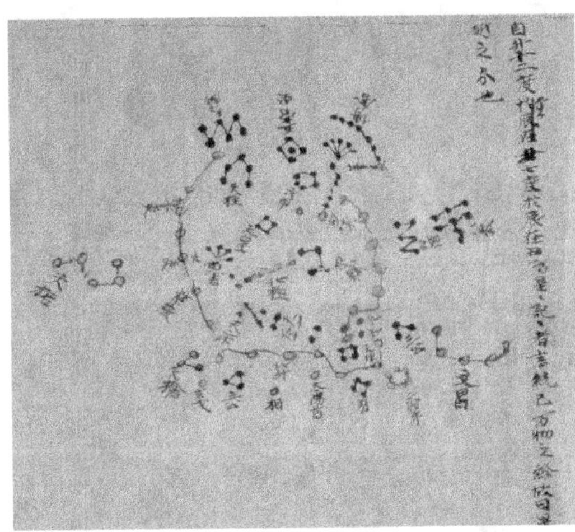

✦ NIGHT SKY MAPPING

Stepping out from the glare of the city and into the embrace of a genuinely dark night sky reveals a celestial display that instantly connects us to the age-old interest our predecessors had with the stars. With a plethora of shimmering lights, the large canvas overhead takes the form of a patterned silver canopy that seems to have a deeper meaning and purpose.

Captivated by the enigma of the night sky, ancient astronomers set out on a profound mission to record, categorize, and interpret the celestial ballet taking place overhead thousands of years ago. Once thought of as far-off and mysterious suns, the stars are now the focus of careful observation, recording, and naming. During this celestial exploration, the oldest star chart—dating back more than thirty thousand years—was discovered. The elaborate design on this ancient relic, which was carved onto an ivory mammoth's tusk found in Germany, strikingly mimics the constellation that is now known as Orion.

Exploring France's underground caverns has revealed incredible historical secrets. Discovered on the stony canvases of these ancient sanctuaries, cave paintings reveal the earliest attempts by humans to chart the night sky. Our ancestors had a basic but deep conversation with the celestial vastness tens of thousands of years before the great civilizations of antiquity embarked on their cosmic excursions.

For these early skywatchers, the night sky, with its enchanting attraction, was both muse and canvas. The basis for the complex tapestry of astronomical knowledge that would emerge in the millennia to come was laid by their crude instruments and astute observations. Our forefathers set out on an everlasting cosmic exploration adventure, identifying constellations and weaving stories into the glittering lights. This journey allowed them to go beyond the confines of Earth and establish a connection with the great secrets of the cosmos.

Among the first ancient societies to map the night sky and give names to some of the stars they saw were the Egyptians. They noted the names of constellations and dubbed the North Star the "star that cannot perish." By recording these early names and patterns and producing astronomical catalogues that named and categorized stars in ever-increasing complexity, the Sumerians and Babylonians went one step farther. Astronomers from Greece, China, and Islam all proceeded to construct increasingly intricate categorization systems; in fact, several stars are still known by their Arabic names today.

The starry background appeared deceptively permanent to the ancients, which perhaps inspired them to document and mythologize the patterns they observed. However, for the first time in history, a specific kind of transient addition to the night sky's lights was seen and recorded in AD 185. Eventually, we were able to recount the tale of the stars' births and deaths by comprehending the nature of this uncommon and amazing phenomenon, which went beyond simply identifying the stars.

The remnants of the cosmic event of AD 185, which blazed over Chinese astronomers' imaginations and skies nearly two millennia ago, were discovered in late 2006. The object known as RCW 86 is seen in the image above, which was captured by the Chandra X-ray Observatory. This object is believed to be the still-glowing remnants of a supernova explosion, one of the most potent occurrences in the cosmos.

The luminous spectacle of a supernova, an awe-inspiring cosmic event, is the grand finale in the life of a massive star. The radiant brilliance of a single star can surpass that of a billion suns during this cataclysmic occurrence. Chinese scientists documented a celestial phenomenon, possibly the remnants of the AD 185 supernova known as RCW 86, which exhibited a dazzling glow for a remarkable eight months before fading from celestial view. Positioned at a staggering distance of approximately 8,000 light years, the sheer luminosity of this distant event is a testament to the immense energy unleashed during the stellar death throes. In their unwitting observation, ancient astronomers unwittingly contributed to humanity's understanding of the inevitable fate of all stars—a poignant reminder that even the most luminous entities in our cosmos are bound by the inexorable law of mortality.

Deep understanding of the nature of our world can be gained from the celestial tapestry, which is woven with the ethereal threads of dying stars. A supernova such as RCW 86 can be seen to have a luminous aftermath that echoes the inevitable death of stars, like a moving cosmic elegy. The first concrete evidence of the finite nature of stars was inadvertently recorded by ancient astronomers as they saw the far-off brightness of the AD 185 supernova. This hushed revelation, echoing through 8,000 light-years and eons, emphasizes the fleeting beauty of celestial bodies and serves as a reminder that even the brightest stars must eventually perish in the celestial cycle of life and death.

✦ STAR NURSERIES

Above us, in the vast starry nurseries, begins the cosmic story of life and death. This is the enormous cosmic play. Nebulae, or these celestial cradles, are rich environments where the cosmic spectacle of newly formed stars takes centre stage. Of all the wonders of the sky, the Orion Nebula is one of the most fascinating characters in this astronomical tale. Located at the far right of the cosmic tableau, it is considered one of the most studied celestial bodies, offering an engrossing look at the complexities of star formation.

The Orion Nebula story begins in 1610, when **Nicolas-Claude Fabri de Peiresc** is commonly cited as the discovery's author. But there's more to this celestial beauty than meets the eye: Mayan folktales reveal that the faint smudge underneath Orion's belt of stars may have been recognized to the ancient Mayans. The ethereal splendour of the nebula is not limited to historical accounts; it is still visible today under extremely clear and dark skies. This intricate and dynamic structure has developed into an astronomical lighthouse, enhancing our understanding of the process of star creation. With its mysterious dance of gas and dust, the Orion Nebula is still a celestial laboratory that helps us understand the origins of stars and makes a major contribution to our understanding of the cosmos.

Hundreds of brilliant young stars brighten the massive interstellar cloud known as the Omega Nebula, also known as the Horseshoe or Swan Nebula. It spans more than fifteen light years. Depending on their mass, these stars may burn for hundreds of millions or billions of years, illuminating the cosmos with a steady stream of light until their insatiable hunger consumes all of the hydrogen in their cores, forcing them to enlarge and become giants.

The most massive stars' grand finale in the cosmic fabric of stellar development is a stunning transition into gargantuan giants. One of these astronomical monsters is the red Mira, a star whose size is 400 times larger than our sun and whose majesty is truly amazing. The red Mira, massive as it is, hovers on the brink of existence, barely hanging on as it approaches the end of its cosmic journey.

The supernova explosion is a historic occurrence that these enormous stars will experience as they near the end of their existence. This apocalyptic event, like celestial fireworks show, signifies the end of a stellar epic and leaves just a faint remnant of the once-powerful star. The formation of a black hole—an intriguing and breathtaking object so dense that not even light can escape its gravitational embrace—is the culmination of the cosmic ballet's most enormous lights.

Stars of somewhat lesser sizes have a different cosmic fate after a supernova. These post-supernova survivors develop into neutron stars, which are celestial remains that radiate radio waves in a characteristic lighthouse beam. These neutron stars, which spin once every few seconds or less, act as cosmic lights and messages, the rhythmic echoes of the powerful stellar events that moulded their existence. Our knowledge of the complex dance of celestial bodies in the enormous cosmic stage is profoundly expanded by the cosmic narrative of these giant stars, from their magnificent ascent to their catastrophic demise.

Stars that are considerably smaller than Mira won't explode. The most prevalent kind of stars in our galaxy are these comparatively cool stars, sometimes known as red dwarfs. Gliese 581 is arguably the most well-known of them that we have examined. This star, which is just over twenty light years from Earth, has attracted a lot of attention lately because at least six exoplanets have been found to be in its orbit. The most intriguing fact is that planet Gliese 581 g is regarded as a top candidate for the hunt for extraterrestrial life since it is believed to orbit within the star's habitable zone.

✦ INTRODUCTION TO EXOPLANETS

The search for exoplanets, or planets outside of our solar system, is one of the most exciting areas of modern astronomical investigation. The fascination and enthusiasm in this discipline are fuelled by the idea of discovering celestial bodies that might be home to extraterrestrial life. Due to planets' faintness and the great distances separating them from our telescopes, the possibility of conducting such a search appeared unfeasible until recently. But today, thanks to state-of-the-art equipment, astronomers use two main methods to reveal the mysterious existence of exoplanets: the transit method and the radial velocity approach.

These methods have revealed a multitude of individual planets and complete planetary systems surrounding hundreds of stars in the delicate ballet of cosmic discovery. These extraordinary objects outside our solar system have masses that vary greatly, from many times Earth's mass to the enormous bulk of 25 Jupiter's. The distance of these recently discovered worlds from their parent stars is a critical factor in determining their potential habitability. Every star system has a **habitable zone**, an area with temperatures suitable for liquid water to exist, which is essential for supporting life as we know it.

The habitable zone's dimensions are directly correlated with the host star's energy output; fainter stars have habitable zones that are closer together and smaller. The red dwarf Gliese 581, which is thought to have at least one planet within its habitable zone, is notable among the cosmic contenders. The promise of discovering far-off worlds with conditions favourable to life becomes a practical reality as astronomers solve the mysteries of exoplanets and habitable zones. This broadens our cosmic perspective and highlights the many wonders that remain to be discovered in the immense expanse of the universe.

Every cosmic structure that we can see in such exquisite detail around us reveals a different aspect of the star's life cycle. But comprehending the existence of stars teaches us something far deeper: they are the ultimate source of all but the most basic of Leucippus' and Democritus' long-sought-after atoms, and as such, they are the building blocks of ourselves. We need to take a brief detour from the skies and firmly return to Earth in order to understand how the stars might be so important to our existence.

⊥ ORIGIN OF LIFE

Taking a close look at the components that make up our existence is the first step in taking the deep trip to understand its origins. The gorgeous Himalayas, which are tucked away in the shadow of the world's tallest mountain range, provide an inspiring and educational backdrop for our exploration. This country of giants is home to nearly 100 peaks that rise beyond the startling height of 7,200 meters (23,620 feet). Nine of the ten tallest mountains on Earth are found here, while the wider Himalayan region is home to 45 of the world's top 50 highest summits.

With their breathtaking grandeur, the Himalayas offer more than simply a gorgeous setting; they also hide an engrossing story that gradually reveals itself as an intriguing beginning towards comprehending the underlying components of the universe. A few tens of millions of years ago, the Himalayas were something very different, despite their current majesty. These massive summits, now immobile in space, conceal the geological mysteries of evolution and change. The sheer enormity of these mountains inspires us to delve into their prehistoric past and investigate the dynamic processes that have created not only the Himalayas but also the very elements that make up the fabric of our cosmos. With its majesty and life-changing past, the Himalayan environment invites us to consider how our own lives are interwoven with the cosmic dramas that have played out over millennia.

The Himalayas are not just the planet's greatest mountain range but also one of the youngest geological marvels. These enormous peaks were non-existent just 70 million years ago, a blip in geological time. The unrelenting dance of Earth's tectonic plates—a dynamic choreography that sculpted the Himalayas with amazing speed—is closely linked to the unfolding drama of their genesis.

The Indo-Australian plate and the Eurasian plate collided at a pace of about 15 centimetres (6 inches) every year throughout this geological period. As a result of this massive collision, the ocean floor between these tectonic rivals crumpled and shifted, creating the magnificent Himalayan Mountain range. Given that the very base of these massive peaks is made up of rocks that were formerly on the ocean floor, it is evidence of the Earth's constant motion. These rocks rose from the depths of an ancient ocean to the lofty heights they currently occupy in a matter of just million years—a blip in the vast tapestry of geological time.

Thus, the Himalayas show not only the workings of geological forces but also the dynamic character of the Earth, where the rocks themselves under our feet are living examples of how tectonic processes can reshape the planet. This story of creation and elevation prompts reflection on the fleeting nature of the geological events that sculpt our globe, leaving a lasting impression on the terrain and advancing our knowledge of how Earth's history is interwoven with a broader cosmic framework.

A careful examination of the Himalayas' geological features reveals solid evidence for their amazing trip. Upon closer inspection, an apparently ordinary chunk of Himalayan limestone shows a powdery, granular structure that provides an intriguing story. What is visible is nothing less than the preserved remains of extinct marine life—once colourful bodies and coral and polyp shells that were abundant in a long-gone ocean millions of years ago. These biological relics go through a metamorphic transition, hardening into the bedrock that makes up the Himalayan landscape through the alchemy of time and a little geological pressure.

With its sea origins that have been petrified, limestone is a visible example of the transformational power of geological processes.

Although the direct precipitation of calcium carbonate from water can also produce limestone, the Himalayas' predominance of the biological sedimentary form highlights the region's rich past entwined with the prehistoric life that once flourished in its waters. The convincing evidence supporting the Himalayan limestone's largely biological composition comes from the amazing

finding of fossils at the peak of Mount Everest. In addition to demonstrating the persistent trace of ancient life on the geological canvas, this discovery serves as a sobering reminder of the Earth's continuous cycle of resource recycling, which has been underway since the planet's birth some five billion years ago.

With its fossils imbedded in it, the Himalayan limestone acts as a geological record, telling the complex tale of the planet's dynamic evolution and ongoing resource recycling. It encourages reflection on the connections between life, geological processes, and the constantly shifting environments that have moulded our world over generations.

Humans are essential participants in this vast cosmic symphony due to the complex interactions between the Earth's biological and geological processes. The fundamental fact is that every atom that makes up our body was once a vital part of something else, which may be unsettling. These atoms were, reassuringly, surely part of the makeup of solid rock, and they may have been involved in the building of an ancient tree or in the development of a dinosaur. The basic simplicity of the universe's building blocks is the basis for the intriguing and seemingly paradoxical phenomenon of rocks changing into live beings and living things going back into rocks.

The basis of this cosmic recycling process is the elemental unity that permeates the universe. These basic components make up the very essence of all living things, including ourselves and inanimate objects. There's a complex dance going on at the atomic level where matter changes states without any noticeable disruption—rocks become life, and life eventually returns to rocks. This recurrent interaction emphasizes how everything is interrelated and how brittle the lines are that separate the living from the lifeless.

This all-encompassing fact encourages reflection on the continuum of existence, in which the atoms that make up our bodies have undergone remarkable transformations as they have travelled through the ages of Earth's history. It emphasizes that, at our core, we are stardust—a harmonic mixture of elements that have cycled through the cosmic tapestry, bridging the gap between the living and the geological—and cultivates a sense of oneness with the cosmos by highlighting how we are one with the universe.

✛ PERIODIC TABLE

In addition to being a list of chemical elements, the Periodic Table is an iconic chart that embodies a long history of scientific discovery and inventiveness. It is frequently associated with science labs in schools. Essentially, this table is a list of the fundamental particles of matter, which are thought to be the smallest units of matter in the universe. Although elemental ideas of matter date back to ancient Greece, the ever-growing list of known elements lacked order until March 6, 1869, thanks to the work of Russian chemist **Dmitri Mendeleev**.

Mendeleev's genius was in his methodical table organization of the sixty-six elements that were known at the time, which was based on their chemical characteristics. This novel arrangement produced predictions for eight as-yet-undiscovered elements in addition to a systematic arrangement of the elements. Surprisingly, during the next thirty years, all eight elements were discovered, and germanium and gallium serve as reminders of Mendeleev's table's predictiveness. The list of known elements grew as scientific research progressed, and by 1955, there were 101. The one-hundred-and-first element was found by scientists at the University of California, Berkeley, and dubbed Mendelevium after the creator of the table.

With 118 elements currently classified, the periodic table has changed over time due to the unrelenting quest of knowledge. In April 2010, a joint Russian-US team successfully synthesized and discovered the most recent addition, ununseptium. This continued development emphasizes the breadth of human knowledge on the constituents of matter and the predictive potential ingrained in Mendeleev's timeless invention. Uncovering the secrets of the elements and offering a fundamental framework for the investigation of the chemical universe, the Periodic Table is a monument to the ceaseless curiosity and cooperative efforts that propel scientific advancement.

The first ninety-four elements of the Periodic Table, which range from plutonium to hydrogen and are all naturally occurring on Earth, are the building blocks of our material existence. These elements shape the complexities of biology and chemistry and are the basic building blocks of the environment we live in. With the exception of plutonium, the other twenty-four elements are synthetic and have brief lifetimes.

It is amazing how the first 94 elements may help one understand biology and chemistry without having to go into the minute details of protons, neutrons, electrons, and quarks' underlying structures. These components are the fundamental building blocks of the natural universe because of their consistency and stability. Extremely high temperatures and energies are needed to break these atoms apart, and these elements are naturally found deep within the blazing furnaces of stars.

The first step in comprehending our cosmic ancestry is our quest to understand the origins of these substances. But before we can begin this journey, we must first realize a startling realization: all that we see above is made of the same fundamental "stuff" that makes up the earth underneath our feet. We are invited to explore the mysteries of our common cosmic ancestry by the universal constancy of matter across the universe,

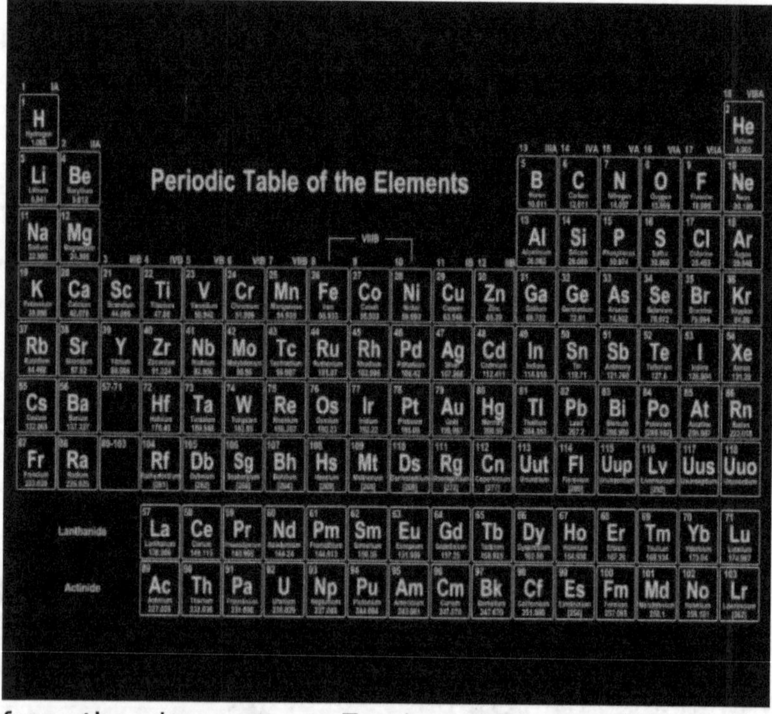

from the elements on Earth to the celestial bodies in the sky. This cosmic symphony highlights how everything is interrelated and provides the framework for an engrossing quest to discover our origins and our role in the enormous cosmic fabric.

It may surprise you to learn that, although humans have only visited one other location in the universe, we are aware of the composition of every star, planet, and moon that can be observed.

When Neil Armstrong and Buzz Aldrin stepped foot on the Moon on July 21, 1969, history was made. It was the first-time humans had ever set foot on another celestial body. Their famous moonwalk lasted for two hours, thirty-six minutes, and forty seconds, but in the final thirty minutes, they completed an important scientific experiment. Buzz Aldrin launched two core tubes into the lunar

surface using simple geological instruments as part of a vital mission to gather rock samples that would go on to become some of the most well-known in history.

Aldrin and Armstrong secured the valuable specimens inside the lunar module while meticulously gathering a total of 22 kilos (47 pounds) of lunar riches. With the aid of a pulley system, they hoisted their scientifically significant payload aboard, shutting the hatch and retiring to rest with their remarkable lunar haul. The lunar rocks that the astronauts slept on became more than simply interesting geological artifacts; they became emblems of a historic victory for science and politics.

Not only was the Apollo 11 mission a huge accomplishment for the United States, but it was also a victory for all of humanity. People all throughout the world were inspired by this unique political success, which transcended national lines and brought people together in awe of the human spirit's power for exploration and discovery. That was a unique moment in human history when political triumph and scientific progress came together to create a lasting impression on the world.

Unknown to most people, a Soviet spacecraft was also in lunar orbit when the Apollo 11 lunar module landed on the moon. The Soviets made three attempts to land on the Moon and gather samples of lunar rock, the most recent being the unmanned Luna 15. Luna 15 was a last-ditch effort to win the race against time in science to return rock samples from another planet. It was launched three days before Apollo 11. Regretfully, Luna 15 did not make it to the Moon's surface in time, despite having successfully started its fall. The moon rocks that Apollo 11 brought back are still being studied today in the ultra-secure laboratories of the Houston, Texas, Lunar Sample Building.

One thing has been evident almost from the beginning, despite forty years of research: these priceless specimens of extraterrestrial geology bear striking similarities to rocks on Earth. The common rock-forming elements oxygen, silicon, magnesium, iron, calcium, and aluminium make up the majority of them, yet nothing on the surface of the Moon is unique from anything on Earth.

Since the historic triumph of Apollo 11, space exploration has revealed a tapestry of cosmic compositions and landscapes that mirror features of our planet. It is evident from landing on Mars and Venus, diving into Jupiter's atmosphere, setting foot on Titan, the moon of Saturn, and traveling to asteroids and comets that the Solar System and our planet have similar elemental ancestry.

Scientists have explored the planet's geology through eight successful landings, and their findings have shown a terrain rich in iron, which has oxidized to give Mars its characteristic rusty red colour. The somewhat alkaline Martian soil contains components such as potassium, sodium, magnesium, and chloride. Sulphur is a characteristic of Venus's thick atmosphere, whereas Mercury is an iron-rich, dense ball with a thin crust mainly made of silicon.

The discoveries keep coming even in the farthest regions of the Solar System, where Neptune is located billions of kilometres distant. It is discovered that Neptune is abundant in organic molecules like methane, which are similar to materials that are widely found on Earth. These findings support the hypothesis that the components of our solar system are uniform throughout its varied celestial bodies.

It is not unexpected from a scientific perspective that our solar system does not contain any new elements. Created more than a century ago, Mendeleev's periodic table offers a thorough structure that can handle every element that is known to exist. But for those who are exploring space, every new landing is a chance to see the familiar components organized in interesting and unexpected ways. Even though physics requires that there be no completely new elements, the explorer's point of view confirms the wonders that arise when seeing the elements that are known to exist in the various environments of other worlds. Seeing these recognizable components in unexpected settings during this cosmic journey is proof of the never-ending wonder of exploration and discovery.

It is reasonable to wonder about the elements that make up the far-off universe and whether the laws of physics can vary in far-off cosmic areas. Even though theoretical physics makes a strong case, it is crucial to avoid assuming universal consistency in the absence of empirical confirmation. The most effective way to verify reality is through experimentation and observation. For stars that are billions of light years away, it may be impossible to provide a clear answer, making them appear untouchable. But our understanding of the makeup of stars precedes our ability to engage with them directly, highlighting the amazing power of astronomical measurements to uncover the fundamental elements of the universe. The search for knowledge expands beyond our immediate surroundings as we consider the grandeur of the universe, stretching the boundaries of discovery to unveil the secrets that lie in the far reaches of the cosmos.

✦ WHAT CONTAINS STARS?

The Sun is located at a great distance of 150 million kilometres (93 million miles) from Earth, making it the centre of our solar system. To get to Proxima Centauri, the nearest star to our solar neighbourhood, one must make the incredible trek of forty thousand billion kilometres, or twenty-five thousand billion miles, over a period of four light years. Proxima Centauri is a red dwarf star that Robert Innes discovered in 1915 at the Cape Observatory in South Africa. It is thought to be a member of a triple star system with the nearby binary stars Alpha Centauri A and B.

Astronomers have learned important lessons from these stars despite their great distances and the fact that our interactions with them are restricted to the light that travels through space to reach us. Over the past century, Proxima Centauri's mass, diameter, and brightness have been carefully measured, providing information about its position within the stellar trio. Beyond the constraints of physical proximity, astronomers have gone deeper into knowing the precise components of stars than just listing important numbers. This is made possible by an intriguing characteristic of the elements: the light that surrounds Earth contains their distinct characteristics inscribed in it.

By analysing the light that stars emit or absorb, scientists can learn a great deal about the chemical makeup of stars through the study of spectroscopy. Since each star's light contains the secret to understanding the elements, every star that can be seen in the sky turns into a cosmic laboratory. Thanks to this sophisticated technique, astronomers may not only determine the makeup of nearby stars but can also reach farther into the universe and learn a great deal about the components that make up the stars that make up the celestial tapestry. We can gain a greater knowledge of the elemental harmony that permeates the cosmos by deciphering the mysteries of faraway stars.

The fascinating tale of using starlight to uncover a star's past dates back to Isaac Newton's seminal study from 1670. The fascinating discovery that light is made up of a spectrum of colours was revealed by Newton in his groundbreaking **Theory of Colour**. He showed how to divide the Sun's white light into its colourful constituents with the straightforward use of a glass prism. A significant discovery regarding the solar spectrum was discovered by German scientist Joseph von Fraunhofer nearly 150 years later, during the calibration of his advanced telescopic lenses and prisms.

In the course of this calibration, Fraunhofer saw something unexpected in the solar spectrum: 574 black lines, which represent several colour-missing or gap-filled areas in the Sun's radiation. Fraunhofer carefully charted the locations of these dark lines, even though he was not at first aware of the full significance of this discovery. He was shocked to see later that the light from the Moon, planets, and other stars contained identical black lines. These characteristic lines, which are now known as Fraunhofer lines, became important landmarks on the path to revealing the cosmic history contained within starlight. This groundbreaking discovery ushered in a new era in space exploration by providing insight into the elemental makeup of far-off celestial bodies and laying the groundwork for spectroscopy.

These sentences were finally given meaning by additional research conducted by two more of the greatest German scientists of the nineteenth century, Robert Bunsen (who is probably best known to schoolchildren worldwide for creating the Bunsen burner) and Gustav Kirchhoff. They made the accurate assumption that these dark spectral lines represented the chemical element fingerprints found in the Sun's atmosphere. Our star's light had travelled 150 million kilometres (93 million miles) across space to bring its constituents' signatures to us.

Kirchhoff and Bunsen's discovery was entirely empirical; they had noticed that heated gases on Earth emit light of very particular colours, and interestingly, those colours depend only on the chemical composition of the gas and not on temperature. Instead of glowing like a piece of hot metal. Specifically, every element in the chemical world has an own colour palette. For instance, the elements strontium, sodium, and copper burn with striking red, deep yellow, and emerald green hues, respectively.

The two German scientists also saw that the elements' blazing colours precisely matched the missing black lines in the solar spectrum. For instance, the two distinct yellow emission lines of heated sodium vapour are precisely matched by two black lines in the yellow portion of the Sun's radiation. This combination of two very slightly distinct yellows is probably recognizable to you; it's the colour of sodium streetlights.

Even though Kirchhoff and Bunsen did not initially know the underlying mechanisms, their partnership in studying the spectral signatures of elements revealed an important feature of nature's activity. Their main objective was to compare the spectral signatures of elements detected on Earth with those of the Sun and stars. The perplexing behaviour of elements, emitting and absorbing light, remained a conundrum until the start of the twentieth century when quantum mechanics supplied an explanation.

The understanding that electrons, which envelop atomic nuclei, cannot exist in random orbits like planets around a star was the fundamental insight that is essential to quantum theory. Rather, they occupy particular, extremely limited "orbits." This limitation results from the fact that electrons are dual entities, displaying characteristics of both waves and particles. Because electrons are essentially waves, their orbits become quantized as a result of being confined less tightly around the atomic nucleus. Electrons can only exist in discrete, quantized energy levels, which is a fundamental conclusion.

In the microscopic world, an electron in an atom moves to a new, more energetic orbit when it absorbs light. On the other hand, light is released when an electron moves from a higher energy orbit to a lower energy orbit. The energy differential between these two orbits must exactly equal the energy of the light that is absorbed or emitted, and this is a crucial criterion. This quantum-mechanical interpretation revealed the complex electron dance inside atoms, connecting light emission and absorption to particular energy transitions. Pioneers such as Kirchhoff and Bunsen's spectrographic work was crucial in driving the development of quantum theory and was a paradigm shift in our understanding of the underlying nature of matter and light at the microscopic level.

The dual nature of light is shown in the field of quantum theory, where it can exist as a stream of particles called photons or as a wave, just like electrons. The most important thing to understand is that every photon has a unique energy that is associated with a specific hue of light. For instance, the energy of red photons is less than that of yellow photons, and the energy of yellow photons is less than that of blue photons. The consequence is profound: every element is selective in what photons it can absorb to move its electrons into higher energy orbits. This is because each element's electrons occupy distinct orbits around the atomic nucleus.

On the other hand, photons of a particular energy are released by electrons when they go from higher to lower energy orbits, giving rise to a very particular colour. This phenomenon explains why, depending on how light it absorbs or emits, each element displays a unique pattern of colours. To put it simply, specific colour absorption and emission indicate the complex atomic structure. A visual representation of the distinct arrangement of electrons surrounding the atomic nucleus is provided by the colours seen throughout these events, which are direct expressions of the energy changes taking place within the atomic structure. This basic idea forms the basis of spectroscopy, which allows scientists to analyse the atomic structure and elemental makeup of celestial bodies, revealing the cosmic symphony contained in the interaction of matter and light.

There are hundreds of Fraunhofer lines in the spectrum of light from our sun, and each one of those lines represents a distinct material in the solar atmosphere that absorbs light as it travels through. The signatures of every element are encoded in the solar code, starting with sodium in the yellow and going all the way through iron, magnesium, and finally to the so-called hydrogen alpha line through red.

Thus, you may determine precisely which elements are present in the Sun by closely examining these lines. It turns discovered that this is roughly composed of 70% hydrogen, 28% helium, and the remaining 2% other elements.

It is important to emphasize that this theory may be used to estimate the components of any star's atmosphere with remarkably high accuracy, not just the Sun but any star visible in the sky. Isn't it amazing that we can tell what those flaming worlds far away are made of merely by observing the light from those glittering stars?

Our scientific intuition that we always observe the signatures of the set of ninety-four naturally occurring elements that we have gathered and identified here on Earth has been validated by these spectrographic studies of the light from space.

Since we are all composed of the same material, it follows that we are inextricably linked to the entire universe, including its billions of stars spread across billions of galaxies. And there's a very straightforward explanation for that, which we shall reveal: everything in the universe has a single genesis.

✛ OUR EARLY UNIVERSE

The early universe, often referred to as the primordial universe, represents the cosmic epoch that unfolded shortly after the Big Bang, the event that marked the birth of the universe. Understanding the early universe requires delving into the first moments and epochs following the singularity, a point of infinite density and temperature from which the universe emerged. The current model of the early universe is based on the Big Bang theory, which posits that the universe began as an extremely hot and dense state, rapidly expanding and cooling over time.

In the first moments after the Big Bang, the universe underwent a period of cosmic inflation, a hypothetical phase during which the universe expanded at an exponential rate. This rapid expansion is believed to have homogenized and smoothed out the distribution of matter and energy, providing a solution to some cosmological puzzles, such as the horizon problem and the flatness problem.

As the universe continued to expand and cool, it entered a phase known as the quark epoch, lasting from about 10^{-12} to 10^{-6} seconds after the Big Bang. During this epoch, the universe was filled with a hot, dense soup of quarks, gluons, and other elementary particles. As it expanded further and cooled, quarks combined to form protons and neutrons, marking the transition to the hadron epoch.

The hadron epoch extended until about 1 second after the Big Bang, characterized by the dominance of hadrons (particles like protons and neutrons) over quarks and gluons. During this epoch, nucleosynthesis occurred, leading to the formation of light elements like hydrogen, helium, and trace amounts of deuterium and lithium.

Following the hadron epoch, the universe entered the lepton epoch, during which leptons (such as electrons and neutrinos) played a significant role. As the universe continued to expand and cool, it eventually entered the era of photon decoupling, around 380,000 years after the Big Bang. During this phase, the universe had cooled enough for electrons to combine with protons and form neutral hydrogen atoms. This decoupling of charged particles from photons allowed the latter to travel freely through space, creating the cosmic microwave background radiation (CMB), which we can still detect today as a faint glow permeating the universe.

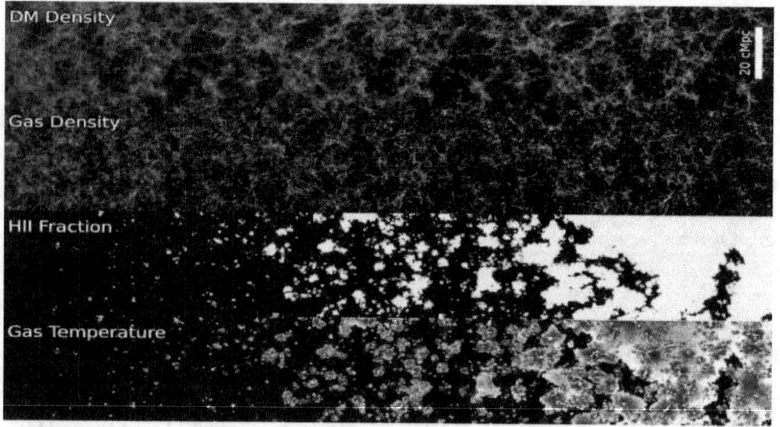

The subsequent evolution of the universe involves the formation of structures like galaxies and stars, marking the transition from the early, hot, and dense stages to the more familiar cosmos we observe today. The study of the early universe provides crucial insights into the fundamental processes and conditions that shaped the cosmos on its journey from the fiery aftermath of the Big Bang to the complex and diverse structure we observe today.

✛ THE BIG BANG

Thirteen billion years ago, the universe began its existence with an event known as the Big Bang. The reasons behind the initiation of the Big Bang and the specific form it initially took remain unsolved mysteries, adding an element of excitement to the field of fundamental physics. The first scientific milestone in the chronology of the universe is identified as the Planck Era, occurring a staggering 10^{-43} seconds after the Big Bang. To grasp the brevity of this period, it is equivalent to 0.0000000000000000000000 0000000000000000001 seconds when expressed in full.

This infinitesimally short duration is intricately tied to the strength of the gravitational force. The exceedingly small value is a consequence of gravity's intrinsic weakness, a phenomenon for which the underlying reason remains unknown. During the Planck Era, the four fundamental forces of nature—gravity, the strong and weak nuclear forces, and electromagnetism—were unified into a singular force, often referred to as a super force. At this point, the universe lacked matter and consisted solely of energy and the super force, presenting a highly symmetric situation in the terminology of physicists. The exploration of this early epoch seeks to unravel the fundamental nature of the cosmos at its inception, providing insights into the unity and subsequent differentiation of the fundamental forces that govern the universe's evolution.

As the universe experienced a rapid expansion and cooling, a series of symmetry-breaking events occurred, shaping the unfolding cosmic drama. At the conclusion of the Planck Era, a pivotal symmetry-breaking event took place, leading to the separation of gravity from the other fundamental forces. This marked the end of perfect symmetry in the early universe. Approximately 10^{-36} seconds after the Big Bang, another significant symmetry-breaking event occurred, signifying the conclusion of the Grand Unification Era.

During this epoch, the strong nuclear force, responsible for binding quarks together within protons and neutrons, diverged from the other forces. This transition was accompanied by an extraordinary period of inflation, a rapid and exponential expansion of the universe. In a remarkably brief span of time—10^(-32) seconds— the universe expanded by a factor of 10^26, an expansion rate equivalent to 100 million million million million times. This inflationary epoch played a crucial role in shaping the large-scale structure of the cosmos.

Notably, this intense period of inflation also marked the entry of sub-atomic particles into the universe for the first time. However, these particles differed from those observed today, as none of them possessed any mass during this early stage. The symmetries breaking during these pivotal moments in the universe's infancy laid the groundwork for the diverse forces and particles that would eventually give rise to the complexity and structure observed in the cosmos today. The exploration of these events provides valuable insights into the fundamental processes that shaped the early universe.

Up until the aforementioned point in the cosmic narrative, the story is theoretically well-motivated but has been relatively untested through experimentation. However, a significant turning point in the cosmic chronicle unfolded approximately 10^(-11) seconds after the Big Bang—an epoch well within our experimental grasp today. This era is actively recreated and observed at CERN's Large Hadron Collider, and it is identified as electroweak symmetry breaking.

During this epoch, the final two fundamental forces of nature—electromagnetism and the weak nuclear force—underwent separation. This process played a pivotal role in endowing the sub-atomic building blocks of all observable matter today, namely quarks and electrons, with mass. The leading theory explaining this phenomenon is known as the Higgs mechanism. The quest for confirming the existence of the associated Higgs Particle stands as one of the paramount objectives of the Large Hadron Collider project.

Theoretical frameworks predict that the Higgs Particle is a key player in this process, interacting with other particles to confer mass upon them. The experimental pursuit of the Higgs Particle represents a crucial avenue for advancing our understanding of the fundamental forces and particles that govern the universe. The ability to probe and scrutinize these high-energy interactions at facilities like the Large Hadron Collider holds the promise of unveiling deeper layers of the cosmos' intricate structure, providing empirical validation for theoretical frameworks and pushing the boundaries of our comprehension of the early universe's dynamic evolution.

The narrative of the universe becomes increasingly grounded in both experimental evidence and robust theoretical frameworks from the epoch approximately 10^{-11} seconds after the Big Bang onwards. At this juncture, our understanding of the universe is substantiated by experiments conducted at particle accelerators, providing a means to validate and verify the underlying physics. The intricate emergence of the familiar particles and forces observed in the contemporary universe is believed to be a consequence of a sequence of symmetry-breaking events initiated at the conclusion of the Planck Era.

The notion of spontaneous symmetry breaking in the early universe aligns with familiar processes observed in phase transitions, such as the transformation from water vapor to liquid water to ice. In these transitions, complex patterns spontaneously manifest as a result of changing conditions, obscuring the underlying symmetry of the initial state. Analogously, the seemingly infinite complexity exhibited by phenomena like snowflakes serves to mask the inherent simplicity of the oxygen and hydrogen atoms forming them. Similarly, the diverse array of forces and sub-atomic particles constituting the building blocks of the contemporary universe conceals the symmetry inherent in the early universe. This symmetry-breaking cascade, initiated in the primordial moments after the Big Bang, has sculpted the cosmic landscape, giving rise to the diverse and intricate structure observed in the universe today. The parallel with familiar phase transitions provides a conceptual bridge for understanding the emergence of complexity from the initially symmetric conditions of the early universe.

The journey toward the formation of protons and neutrons, the fundamental building blocks of elements, and the birth of the first chemical elements began around a millionth of a second after the Big Bang. At this point, the quarks, the sub-atomic particles that make up protons and neutrons, had cooled sufficiently to be bound together by the strong nuclear force. The simplest element, hydrogen, composed of a single proton, made its appearance within this brief cosmic timespan. Just a millionth of a second into the life of the universe, the groundwork for the formation of chemical elements was laid.

After approximately three minutes, the universe had cooled further, allowing protons and neutrons to combine and form helium. Helium, with two protons and one or two neutrons in its nucleus, became the second-simplest chemical element. Small quantities of lithium, with three protons, and beryllium, with four protons, also emerged during this phase, representing the third and fourth simplest elements in the cosmic inventory. Beyond this point, the process largely halted.

By the three-minute mark, the universe had settled into a state where the four fundamental forces we recognize today—gravity, the strong and weak nuclear forces, and electromagnetism—were established. The elemental composition at this stage comprised roughly 75 percent hydrogen (by mass) and 25 percent helium. This marks the culmination of the early universe's journey toward the creation of the simplest chemical elements, encapsulating successive symmetry-breaking events that shaped the evolving cosmic landscape.

✦ SUB-ATOMIC PARTICLES

Sub-atomic particles, the constituents of atoms, play a pivotal role in understanding the fundamental nature of the universe. These particles are incredibly small entities, residing within the nucleus or orbiting around it. Quarks are elementary particles that combine to form protons and neutrons, which are key components of atomic nuclei. Leptons, another category of sub-atomic particles, include electrons, integral to the structure of atoms. Electrons, with a negative charge, orbit the nucleus, creating the electron cloud. Neutrinos, nearly massless and electrically neutral, are elusive particles that interact weakly with matter.

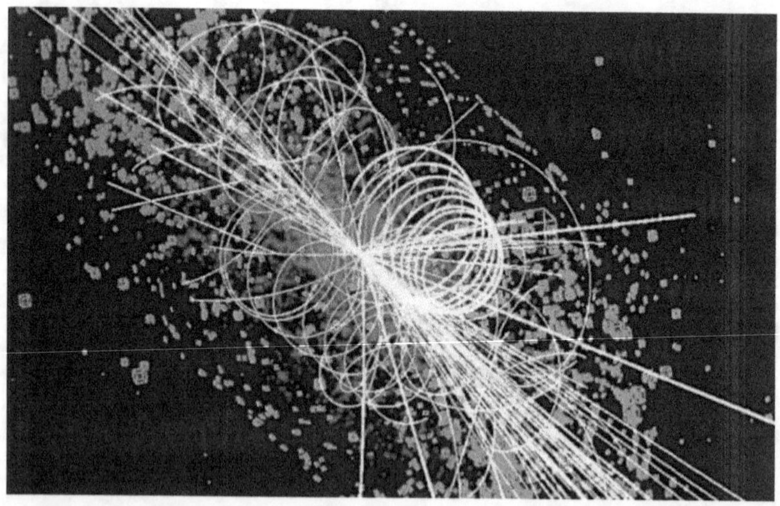

Bosons, the force carriers, mediate interactions between particles. Photons, representing the electromagnetic force, enable light transmission.

Gluons facilitate the strong nuclear force binding quarks together within protons and neutrons. W and Z bosons are responsible for weak nuclear interactions. The Higgs boson, associated with the Higgs field, is crucial for understanding mass acquisition by particles.

Sub-atomic particles exhibit dual wave-particle characteristics, as revealed by quantum mechanics. This inherent duality contributes to the complex behaviour of particles. The study of these particles deepens our comprehension of the fundamental forces governing the universe, unravelling the intricate tapestry of particle physics.

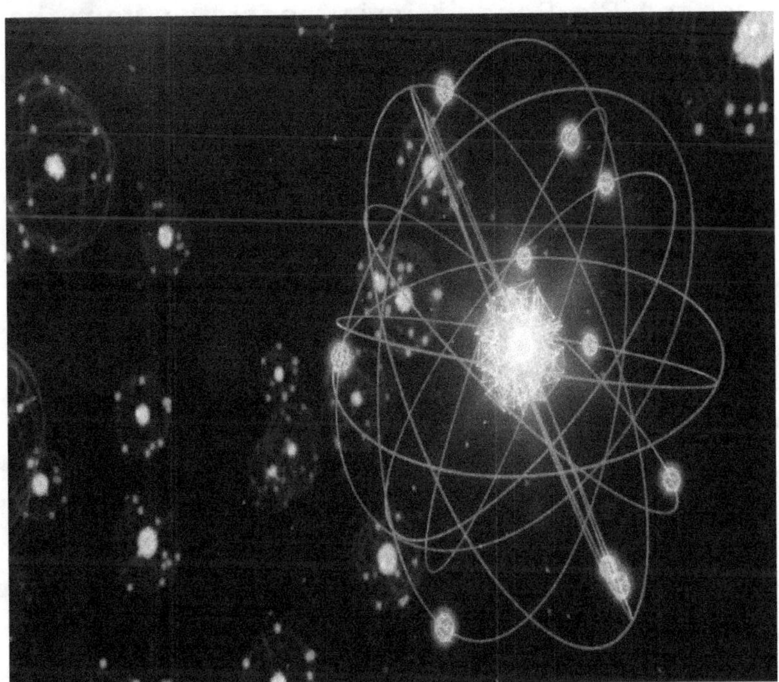

✛ TIMELINE OF UNIVERSE

The timeline of the universe unfolds as a captivating saga of cosmic evolution, tracing its trajectory from the monumental event of the Big Bang to the present day.

1. **Planck Era** (Up to 10^{-43} seconds after the Big Bang): The universe initiates in a singularity, and during the Planck Era, the four fundamental forces—gravity, strong nuclear force, weak nuclear force, and electromagnetism—exist as a singular force. The universe expands exponentially during inflation.

2. **Grand Unification Era** (Up to 10^{-36} seconds after the Big Bang): Gravity separates from the unified force, marking the end of the Planck Era. Subsequently, the strong nuclear force diverges from the other forces. Inflation, an intense expansion, occurs.

3. **Electroweak Symmetry Breaking** (Around 10^{-11} seconds after the Big Bang): Electromagnetism and the weak nuclear force separate, setting the stage for the emergence of particles with mass. This era is experimentally explored at the Large Hadron Collider.

4. **Quark Epoch** (Up to 1 microsecond after the Big Bang): Quarks, fundamental constituents of protons and neutrons, become bound by the

strong nuclear force, forming these building blocks of atomic nuclei. The first chemical element, hydrogen, appears.

5. **Nucleosynthesis** (First few minutes after the Big Bang): Helium and trace amounts of lithium and beryllium form as protons and neutrons combine. The universe, now composed of approximately 75% hydrogen and 25% helium, experiences the formation of the simplest chemical elements.

6. **Dark Ages** (Up to 100 million years after the Big Bang): The universe enters a phase of darkness as cosmic structures evolve, and the first stars and galaxies emerge, ionizing the intergalactic medium.

7. **Cosmic Renaissance** (Around 150 million years after the Big Bang): Light from the first stars and galaxies permeates the universe, ending the cosmic dark ages. The epoch of reionization ensues.

8. **Formation of Cosmic Structures** (Over billions of years): Gravity Molds the cosmic web, leading to the formation of galaxies, galaxy clusters, and cosmic filaments. Stars and planetary systems arise within galaxies.

9. **Formation of Earth and Solar System** (Around 4.6 billion years ago): A protoplanetary disk

coalesces to form the Sun and the solar system. Earth, with its diverse ecosystems, evolves over geological epochs.

10. **Emergence of Life** (Around 3.5 billion years ago): Life originates on Earth, evolving from simple unicellular organisms to complex multicellular life forms.

11. **Human Evolution** (Around 2 million years ago): Homo sapiens, the modern human species, emerges through the process of evolution.

12. **Cultural and Technological Evolution** (Over the last few thousand years): Human civilizations arise, developing agriculture, language, writing, and advanced technologies.

13. **Modern Era** (From the Industrial Revolution to the Present): Rapid advancements in science, technology, and societal structures characterize the modern era, shaping the contemporary world.

The timeline of the universe stands as a testament to the intricate interplay of cosmic forces, leading to the emergence of galaxies, stars, planets, life, and the complexity of the present-day cosmos.

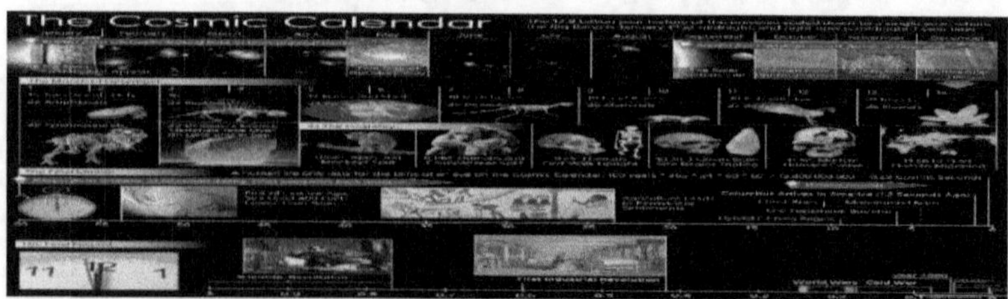

✛ MATTER BY NUMBERS

Throughout human history, the development of civilization has been closely associated with the discovery and application of particular chemical components. It is thought that people began mining and crafting copper 11,000 years ago. The unique properties of this metal marked the beginning of a new era in technology, marked by the switch from stone to metal tools and weapons. It happened again four millennia later, this time with iron, which is still used today to combine with carbon to create steel, the alloy that serves as the framework for industrial civilization.

These two elements' unique physical attributes caused them to play a part in our history. Since copper is one of the few metals that exists naturally in its pure form and is such an unreactive chemical, it was most likely the first metal used by humans. Additionally, it is highly pliable and soft, making it reasonably simple to shape into tools and weapons. Copper and another metallic element, such as tin, combine to make the alloy bronze; copper and zinc combine to form brass. Surprisingly, iron is the most prevalent element on Earth and ranks fourth in the rocks that make up the crust. Despite being more challenging to extract and deal with than bronze, iron is a great material to make weapons out of because it is harder and lasts longer than bronze.

These two metals, which are located just a few places apart in the periodic table, have had a significant impact on human history. Elements 26 and 29 are iron (Fe) and copper (Cu), respectively. Naturally, the earliest people to use these metals would not have understood why the two elements differed and were physically identical. What, therefore, is their primary distinction from one another? The solution is surprisingly easy. Each element's atom is made up of three components, as previously mentioned: protons, neutrons, and electrons. The quarks inside protons and neutrons remain locked away at the temperatures seen on Earth, so we do not need to take them into account. Thus, when talking about the chemistry of Earth, we can ignore them.

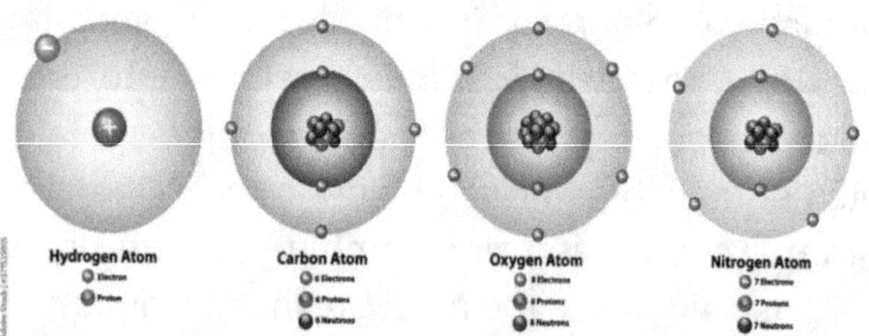

The intricacies of atomic composition reveal a fascinating interplay of sub-atomic particles.

Hydrogen, the simplest element, features an atomic nucleus housing a solitary proton. With a positive electric charge, the proton has the capability to ensnare an electron, which, in turn, carries a negative electric charge. This pairing results in electrically neutral hydrogen atoms.

Delving deeper into the sub-atomic realm, the proton itself is composed of three quarks—two up quarks and one down quark. The up quark bears an electric charge of +2/3, while the down quark possesses a charge of -1/3. The electron, on the other hand, bears a charge of -1. The remarkable aspect lies in the precise combination of these quarks within the proton, resulting in a balanced and neutral overall charge.

The neutron, a partner to the proton within atomic nuclei, is constructed from two down quarks and one up quark. This specific combination imparts a neutron with no electric charge. The symmetry and precision in charge balancing between quarks and electrons constitute a profound mystery, presenting a formidable challenge for physicists in the twenty-first century to unravel and explain this intricate dance of fundamental particles.

The distinctive characteristics of chemical elements stem from the number of protons within their atomic nuclei, a fundamental attribute that defines each element. Notably, the number of neutrons in the nucleus does not influence chemical properties; it is the behaviour of electrons orbiting the nucleus that dictates the chemistry of an element. The equality between the number of protons and electrons ensures electric neutrality in atoms.

Take the example of hydrogen, which consists of one proton and one electron. Another variant of hydrogen, known as deuterium, features a neutron alongside the proton in its nucleus. However, this addition of a neutron doesn't alter its chemical properties, as the crucial factor is the singular electron. Technically, deuterium and hydrogen represent two isotopes of the same element.

Moving to helium, it consistently possesses two protons and two electrons. Helium manifests in different forms, including helium-3 and helium-4, distinguished by the presence of one or two neutrons in the nucleus. The trend continues through the periodic table, with each succeeding element harbouring one additional proton in its nucleus. The role of neutrons is vital in maintaining the nucleus's cohesion through the strong nuclear force. Unlike protons, neutrons lack electric charge, mitigating the electrostatic repulsion that protons would otherwise exert on each other within the nucleus. This phenomenon becomes particularly crucial in heavier nuclei, where the ratio of neutrons to protons tends to increase, contributing to the stability of the nucleus.

Thus, the synthesis of chemical elements is straightforward. Three protons and a few neutrons

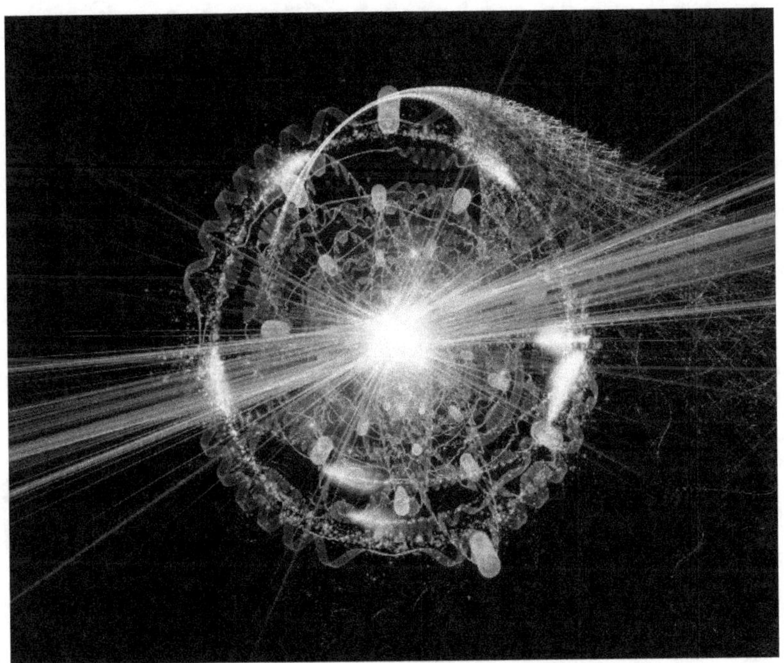

added to the nucleus of iron will transform it into copper. That is the only thing involved. Naturally, this is easier said than done, yet nature is capable of accomplishing it because the first four chemical components were present in the universe within minutes of its creation. The heavier components were assembled later, but the building blocks were already in place.

☩ POWERFUL EXPLOSION ON EARTH

Two of the finest physicists of all time had grown disenchanted with the idea years before the Manhattan Project devised and produced the most devastating weapon ever used in combat. Before any kind of nuclear bomb had been assembled in 1941, the thoughts of friends and coworkers Edward Teller and Enrico Fermi had already strayed beyond the bomb that would later be dropped on Hiroshima and Nagasaki with devastating effect. Both men would go on to be members of the Manhattan team.

Fission bombs, such as the ones dropped on Hiroshima and Nagasaki, work by dividing the nuclei of extremely heavy atoms—plutonium in the case of the Nagasaki bomb, and uranium in the case of the Hiroshima bomb—into lighter elements like caesium and strontium. This is the component assembly done backwards. Neutrons are released when a uranium or plutonium nucleus breaks, which causes the breaking of more nuclei.

A nuclear chain reaction results in this manner. A significant quantity of energy is released when a heavy nucleus divides; this energy is known as "nuclear binding energy" and is kept in reserve by the potent nuclear force field that holds protons and neutrons together inside the nucleus.

But even years before the concept of a fission bomb became a reality, during the very early phases of the Manhattan project, **Enrico Fermi** conjectured that the possibility of developing a much more potent kind of bomb existed. **Edward Teller** developed an obsession with his friend's concept and devoted the following ten years to creating the most potent explosions ever recorded in Earth history. Teller became known as the **father of the hydrogen bomb** as a result.

On November 1, 1952, the collaboration between Enrico Fermi and Edward Teller culminated in the realization of Ivy Mike, the codenamed test for the first successful hydrogen bomb. Conducted on the Eniwetok atoll in the Pacific Ocean, the explosion surpassed all previous magnitudes, estimated to be 450 times more potent than the bomb dropped on Nagasaki. The destructive force generated a colossal fireball, spanning over five kilometres in width, and left in its wake a two-kilometre-wide crater, erasing the tiny atoll from existence. Teller, in partnership with fellow Manhattan Project scientist Stanislaw Ulam, designed this unprecedented bomb, although Teller himself observed the event from the distance of his office in Berkeley, California, using a seismometer. The shockwave's impact was so discernible that, with a touch of cryptic humour, Teller informed his colleagues of the success by declaring, "It's a boy!"

The first nuclear fusion reaction created by humans was the Ivy Mike test. Nuclear fusion is the process by which two atomic nuclei fuse together to form a single heavier element. It is the direct opposite of fission. The process of assembling hydrogen into helium, which took place in the early moments of the universe's evolution, is replicated by the hydrogen bomb.

Today, all five of the major nuclear weapon states use the Teller-Ulam design, which was used for the hydrogen bomb that exploded on Eniwetok. Even though the design's fusion component alone accounts for a small portion of its explosive potential, when paired with the bomb's other stages, it unleashes devastation never seen before.

The processes of fission and fusion, despite being seemingly opposite, both release vast amounts of energy due to the intricate interplay between the electric repulsion of protons in the nucleus and the force of the strong nuclear force binding protons and neutrons together. This delicate balance results from the competition between forces attempting to break the nucleus apart and those striving to hold it together.

In fission, a heavy element is split, and the resulting fragments are generally lighter than the original element. This process is energetically favourable for elements heavier than iron and nickel. Conversely, fusion involves combining lighter elements, releasing energy in the process. Iron and nickel, due to their proximity to the optimal mixture of stability, stand as benchmarks. Elements lighter than iron and nickel can achieve greater stability by fusing, while elements heavier than these can enhance stability by undergoing fission. The net result is the liberation of energy as nature navigates this delicate equilibrium between attractive and repulsive forces within atomic nuclei.

To be absolutely correct, we need point out that the stability of the elements is influenced by more than only the equilibrium between the nuclear and electromagnetic forces. These have to do with the nucleus's physical structure and the fact that protons and neutrons are favoured in balance for quantum mechanical reasons. (If you're curious, enjoy searching for "Semi-empirical mass formula" on Google!)

Fusion may appear to be the pinnacle of human technical advancement on Earth, but in reality, it is the most natural phenomenon on the planet. It's a process that's happening all over the universe right now, not just at the Big Bang. In actuality, it constantly occurs millions of kilometres above our heads, lighting up the entire universe.

The process of fusion drives all-stars, including our sun, in the cosmos. Gazing upon an unfathomably vast expanse of incomprehensible nuclear explosion energy, you are enveloped in it. The Sun is actively converting hydrogen into helium at a rapid pace at its core, 800,000 kilometres (500,000 miles) below the surface, where temperatures can approach fifteen million degrees Celsius. The Sun can change 600 million tonnes of hydrogen into helium in a single second, releasing as much energy as humanity will require over the course of the next million years. This energy is responsible for the stars' brightness and the solar system's heat and light.

All life on Earth is made possible by the process of converting hydrogen into helium, yet despite the Sun's immense power, it can only change hydrogen, the simplest element, into helium, the next simplest. Throughout the night sky, this process is repeated; all stars in the universe were first propelled by this reaction and fuelled by hydrogen.

The genesis of the second-simplest element, helium, is well understood, facilitated by processes occurring in stars, the early Universe, and even replicated on Earth. However, this elucidation doesn't extend to the origin of the remaining ninety-two naturally occurring elements. The prevalence of these elements

throughout the Universe, constituting the very essence of our planet, prompts the search for a prolific source. Our bodies, composed of billions and billions of atoms ranging from magnesium to zinc, iron, and the pivotal element for life, carbon, present a puzzle. Carbon, in particular, is an indispensable building block, with each human being comprising an astronomical number of carbon atoms that did not exist in the early moments of the Universe. The solution lies in the process of nuclear fusion, and the natural arena to explore is within the celestial bodies— the stars themselves.

✛ RED GIANT STARS

It would seem that a star would gradually flicker out of existence as its hydrogen reserves were depleted, but stars like our sun experience the reverse. A star that is running out of hydrogen really swells up to possibly hundreds of times its initial size since the core has been the star's beating heart for millions or billions of years. These stars are referred to as red giants.

Betelgeuse, officially designated Alpha Orionis, stands as one of the closest red giants to Earth, captivating stargazers with its brilliance and distinctive reddish hue. Positioned a mere 500 light years away, it holds a notable place as the ninth-brightest star in our night sky. The star's luminosity has drawn attention throughout history, with Sir John Herschel's nineteenth-century observations documenting its significant brightness variations. However, the revelation of Betelgeuse's extraordinary nature emerged when astronomers from the Mount Wilson Observatory, namely Albert Michelson, Francis Pease, and John Anderson, employed interferometry to measure its diameter. This technique, assessing the angular diameter from Earth's perspective, unveiled Betelgeuse's true magnitude. Approximately twenty times the mass of our sun, Betelgeuse's colossal size is even more striking. Placing it at the centre of our solar system would dwarf our sun, extending beyond Earth's orbit and reaching out to Jupiter. The current estimate pegs Betelgeuse's diameter at around 800 million kilometres (500 million miles), portraying a celestial marvel that would engulf our solar system with its expansive and ethereal presence.

We can study Betelgeuse in great detail because of its enormous size and close closeness. Betelgeuse was the subject of the first direct image of another star to be captured by the Hubble Space Telescope in 1996, revealing its disc and surface characteristics. We've even been able to image sunspots on its surface and gather ever-more detailed information about its atmosphere. But the answer to where the heavy elements are formed is not on the crimson giant's surface; rather, we must travel deep into its dying heart to fully comprehend this.

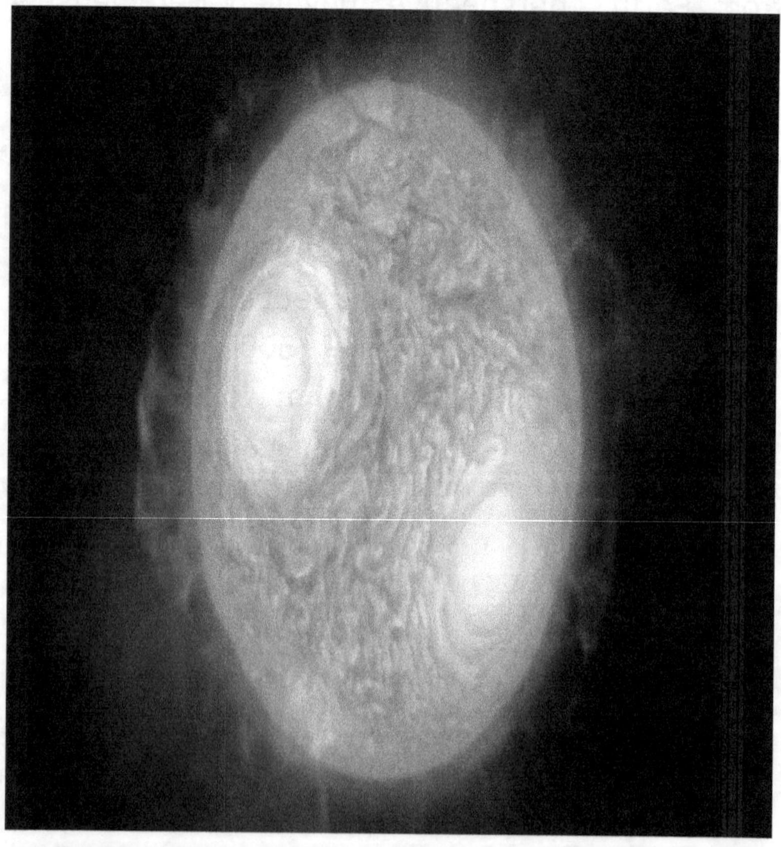

✠ DEATH OF STAR

The building itself stood as a hollowed relic, its original form reduced to a skeletal framework of bricks, devoid of any semblance of windows. The cells within resembled dormitories, featuring rows of twenty or thirty concrete bunk beds arranged closely. These spaces, each equipped with a diminutive bathroom, bore witness to a stark absence of privacy, with tattered cloths still hanging across some entrances. The cell walls presented a bizarre mosaic, a chaotic blend of torn colours, adorned with images ranging from glamour girls to sporadic football team depictions. The scene stirred disquiet for two distinct reasons. Firstly, contemplation arose about the grim prospect of enduring incarceration within this environment—a steel and concrete cage in the heart of Rio's hot and humid city. Secondly, a more visceral unease emerged: the prison was rigged with live explosives. As I navigated through the labyrinthine structure, the contrast between the vibrant external world and the internal abyss became palpable. The radiant exterior pressed against the structure, resembling a stellar surface, yet inaccessible from within the dark confines. The analogy resonated as I descended precarious, cement-dusted stairwells into the profound core of this dying star. Inside this condemned and violent edifice, far removed from the surface's light, the elements crucial to life were meticulously assembled. Here, the star underwent a transformative shift, transitioning from a consumer of matter to a producer of the building blocks of life.

Stars exist in a delicate balance, where gravity seeks to compress them, leading to an increase in temperature. Eventually, this heat overcomes the electromagnetic repulsion between hydrogen atoms, initiating fusion to form helium. The energy released from this process counteracts gravity, maintaining the star's stability. However, as the hydrogen fuel depletes, gravity regains control, causing the core to collapse rapidly, leaving behind a shell of hydrogen and helium. Within the contracting core, temperatures soar to around 100 million degrees Celsius, triggering a new fusion process—helium burning.

This shift from hydrogen to helium fusion has two significant outcomes. First, the released energy halts the collapse, leading to the star's stabilization and rapid expansion, marking the onset of its phase as a red giant. Second, this fusion gives rise to a crucial element for life.

Initially, the fusion of two helium nuclei should produce the unstable isotope beryllium-8, which quickly breaks down. However, in the intense temperatures of a dying star, with the core exceeding 100 million Kelvin, these nuclei persist just long enough to merge with a third helium nucleus, giving birth to the precious element carbon-12. This process is the origin of all carbon in the Universe, with every carbon atom in every living entity on Earth being forged in the heart of a dying star.

The alchemical synthesis of carbon marks the end of the helium-burning phase of the star's life, yet conditions during that same extremely hot period allow a helium nucleus to latch onto a freshly formed carbon nucleus to generate another element essential to life. As the third most abundant element in the universe after hydrogen and helium, oxygen makes up 21% of the air we breathe and is necessary for water, the solvent of life. It's important to keep in mind that the 2.5 grams of oxygen you breathe in every minute came from an environment that is as far removed from what we consider to be habitable as possible.

In the grand timeline of a star's existence, the cosmic assembly line responsible for crafting carbon and oxygen operates swiftly, completing its task within a mere million years. As helium depletes in the star's core, many stars, particularly those of average size like our sun, reach the culmination of their productive phase. When our sun, for instance, approaches this juncture in approximately ten billion years, it won't possess the gravitational energy required to further compress the core and reignite fusion.

The aging star, now increasingly unstable, experiences the accumulation of immense pressure points. Ultimately, the entire stellar atmosphere erupts in a magnificent explosion, propelling its valuable cargo of elements—oxygen, carbon, hydrogen, and more—into the vastness of space. During this ephemeral period, lasting no more than a few tens of thousands of years, a dying star adorns the cosmos with one of its most exquisite creations: a planetary nebula.

In the cosmic aftermath of the dazzling planetary nebula display, an average-sized star embarks on a transformation that culminates in it shrinking to a size no larger than Earth—a state known as a white dwarf. This fate awaits countless stars similar to our sun. However, for the massive luminaries like Betelgeuse, the stellar drama unfolds with further acts. If a star boasts a mass approximately 1.5 times that of our Sun, its journey down the chemical production line persists.

As the fusion of helium gradually concludes, gravitational forces seize control, prompting the re-collapse of the core. The escalating temperature initiates the third phase in the creation of the universe's elements. With temperatures soaring into the hundreds of millions of Kelvins, carbon undergoes fusion with helium, yielding neon. Subsequently, neon fuses with more helium, producing magnesium, while two carbon atoms unite to form sodium. In this cosmic cauldron, as additional elemental ingredients join the mix and temperatures escalate, a succession of heavier elements emerges. The core continues to collapse, temperatures continue to surge, and a new phase of fusion commences, leaving behind layers of freshly forged elements.

After the initial creation of the first twenty-five elements within the star, the stellar production line encounters a significant hurdle at the twenty-sixth element—iron. This element is crafted through a intricate sequence of fusion reactions powered by silicon. At this stage, the star's temperature soars to a minimum of 2.5 billion Kelvin, but it faces an impasse. The pinnacle of nuclear stability is attained, and no additional energy can be released by introducing more protons or neutrons to iron.

The concluding phase of iron production transpires swiftly, spanning a mere couple of days. In a desperate bid to extract every last ounce of nuclear binding energy and thwart the relentless pull of gravity, the star's core undergoes a transformative process, evolving into nearly pure iron. At this  juncture, the fusion process grinds to a halt. Once the star's core has been transmuted into iron, its remaining moments are mere seconds. Gravity now prevails, and the star succumbs to its own gravitational forces, giving rise to the formation of a planetary nebula.

✛ SUPERNOVA

In the celestial drama of a star's life, a pivotal and cataclysmic event is the supernova. When a massive star exhausts its nuclear fuel, particularly iron accumulating at its core, it confronts an inevitable fate—a spectacular and violent end. The balance between the star's gravitational forces pulling inward and the outward pressure generated by nuclear reactions reaches a critical juncture. Unable to sustain this equilibrium, the star implodes under its colossal gravitational pull.

This implosion triggers an extraordinary release of energy, generating shockwaves that reverberate outward. The outer layers of the star are expelled into space, forming a mesmerizing display known as a supernova explosion. The intense brightness during this explosive event can briefly outshine an entire galaxy, making it visible from vast cosmic distances.

As the remnants of the stellar explosion disperse into the cosmos, they carry with them the heavy elements forged within the star's core. These elements, including carbon, oxygen, and metals crucial for the formation of planets and life, are scattered across the interstellar medium. The shockwave initiated by the supernova can also induce the compression of nearby interstellar clouds, potentially triggering the birth of new stars.

The aftermath of a supernova showcases a celestial spectacle—an expanding cloud of gas and dust known as a supernova remnant. Within this turbulent environment, particles are accelerated to tremendous speeds, emitting various forms of radiation. The remnants of the supernova contribute significantly to enriching the interstellar medium with diverse elements that serve as the raw materials for future celestial bodies.

In some instances, the core of the collapsed star transforms into an incredibly dense object—a neutron star or, in more extreme cases, a black hole. These enigmatic remnants, endowed with immense gravitational forces, continue to influence their cosmic surroundings.

The study of supernovae provides astronomers with invaluable insights into the life cycle of stars, the production of elements, and the dynamic processes shaping our universe. As these stellar explosions continue to captivate the astronomical community, they remain pivotal phenomena, offering glimpses into the profound mechanisms governing the cosmos.

✚ BEGINNING AND THE END

The fate of the biggest stars in our cosmos is dramatically different after a few million years of existence. Giant stars teeter on the verge of collapse, having burned through every element but iron after running out of hydrogen. However, despite their deteriorating health, these celebrities have one last violent deed, and it's a kind one. It happens so intensely that the heavy components can form as a result.

In the final act of a dying giant star, a profound and dramatic transformation takes place at its core. As the fusion reactions come to a halt, gravity seizes control, causing a rapid and intense collapse of the core. This implosion occurs at speeds of up to a quarter of the speed of light, leading to a substantial increase in temperature and density. The core, once a massive structure, now shrinks to a fraction of its original size, potentially reaching a diameter as small as 30 kilometres (19 miles).

At this critical juncture, the extreme conditions trigger the onset of quantum mechanics, forcibly merging electrons and protons into neutrons. The Pauli exclusion principle, governing the behaviour of neutrons, asserts itself, preventing them from occupying the same quantum state. Consequently, a ball of neutrons forms, representing the most rigid material in the Universe—100 million million million times harder than a diamond.

This dense core, composed of tightly packed neutrons, reaches a point where further compression becomes impossible. The abrupt cessation of the collapse results in a powerful rebound, propelling a shockwave through the star. As this wave encounters the outer layers, it generates temperatures unparalleled in the Universe—reaching a staggering 100 billion degrees. Although the precise mechanism for this rapid heating is not fully understood, the intense conditions persist for mere seconds, allowing for the formation of the heaviest elements observed in the universe, including gold and plutonium.

This cataclysmic event, known as a Type II supernova, stands as the most potent explosion known to us, unveiling the incredible forces and processes at play within the heart of dying massive stars. The remnants of such explosions contribute significantly to the cosmic enrichment of heavy elements, shaping the celestial landscape with their awe-inspiring aftermath.

Since the dawn of modern science, we have never had the opportunity to witness a supernova up close because they are so rare. A few years before to the development of the astronomical telescope, in 1604, the last supernova explosion in our galaxy that was visible from Earth occurred. The Milky Way is supposed to have one supernova explosion every century on average, yet for the past 400 years, nothing has happened. It's long overdue, and astronomers are always looking up into the sky to find stars that they believe have the highest probability of going supernova.

Betelgeuse, the brilliant red jewel in the constellation Orion, stands out as a prime candidate for a spectacular cosmic event—the eruption of a supernova. Intently observed by numerous telescopes over the years, Betelgeuse's behaviour reveals a high level of instability, with a notable 15% dimming in brightness observed in the past decade. This volatility has positioned Betelgeuse at the forefront of potential supernova candidates.

Being a relatively youthful star, estimated to be around ten million years old, Betelgeuse has rapidly traversed its life cycle due to its massive nature. Despite the astronomical timescales involved, the expectation is that Betelgeuse could go supernova at any moment. The uncertainty surrounding this timeframe indicates that, in stellar terms, "any time soon" encompasses the possibility of an explosion either tomorrow or within the next million years. What is certain is that when Betelgeuse does eventually go supernova, it promises to deliver a captivating spectacle.

Situated a mere 500 light years away, Betelgeuse's proximity ensures that the supernova event will be exceptionally luminous. Anticipated to outshine all other celestial bodies, it might even rival the brightness of a full moon at night and cast a radiant glow akin to a second sun during the day. The impending supernova of Betelgeuse holds the potential to offer a remarkable celestial display for Earth-bound observers.

Upon the eruption of Betelgeuse into a supernova, a momentous release of energy will occur, surpassing the total energy output our sun would generate throughout its entire lifetime. The cataclysmic explosion will rip the star apart, propelling into space all the elements synthesized during its evolutionary journey.

Over the course of millions of years, these newly formed elements will disperse, giving rise to a nebula—a celestial expanse rich in diverse chemicals meandering through the cosmos. At its core, the only remnants will be the super-dense nucleus of neutrons, a product of the gravitational forces that have compressed the once expansive star. This neutron star signifies the ultimate fate of Betelgeuse—an exceedingly dense, intensely hot mass equivalent to the Sun's mass yet condensed into a mere 30 kilometres (19 miles) in diameter.

While direct observations of neutron stars remain elusive, advanced technologies such as X-ray imaging have enabled scientists to gain crucial insights into these enigmatic celestial bodies.

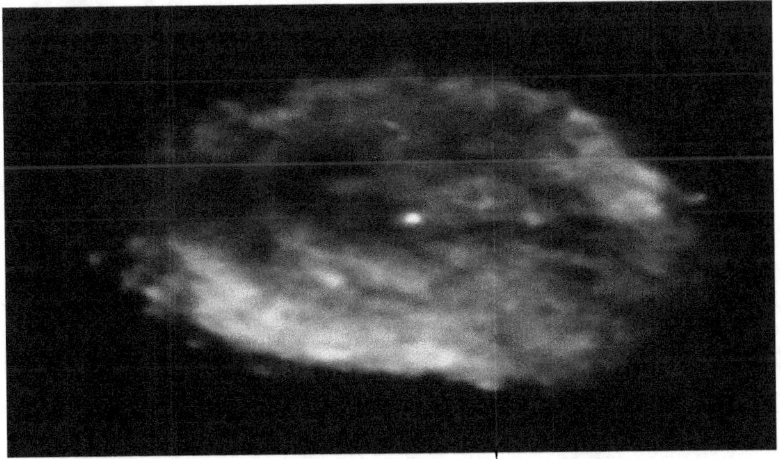

Recent images of RCW103, a remnant from a supernova event transpiring two thousand years ago and located approximately 10,000 light years away, contribute valuable data to our understanding of neutron stars, even from a distance.

The cosmic drama of stellar deaths and rebirths is a grand celestial symphony, echoing the Earthly cycle of life, death, and renewal on an unimaginable cosmic scale. In the constellation of Orion, within the sword handle, lies the Orion Nebula—a spectacle visible to the naked eye as a misty patch of light, but through a telescope, it unfolds as a breathtaking marvel of the Universe. Concealed within its nebulous clouds are luminous points, the nascent stars arising from the remnants of supernova explosions—a testament to the birth of the new amid the demise of the old.

This cosmic ballet mirrors the process that birthed our own solar system. Five billion years ago, our sun emerged from a nebula akin to the one we observe in Orion. Around this radiant star, a cosmic dance unfolded, and from the remnants of past stellar deaths, planets coalesced. Among them, Earth took shape—a celestial body composed of elements forged in the fiery finales of stars, drifting through the cosmos within the remnants of a nebula.

Yet, the narrative doesn't conclude here, as recent scientific conjecture posits that the chemical elements themselves may not be the most intricate components woven together to form the essence of our existence in the cosmic crucible. The journey from stardust to life, from the remnants of ancient stars to the intricate tapestry of life, unfolds as an awe-inspiring odyssey within the vast cosmic theatre.

✛ ORIGIN OF LIFE

The graph, which shows the spectrum of light from the Orion Nebula as seen by the Herschel Space Observatory Telescope, may not look very

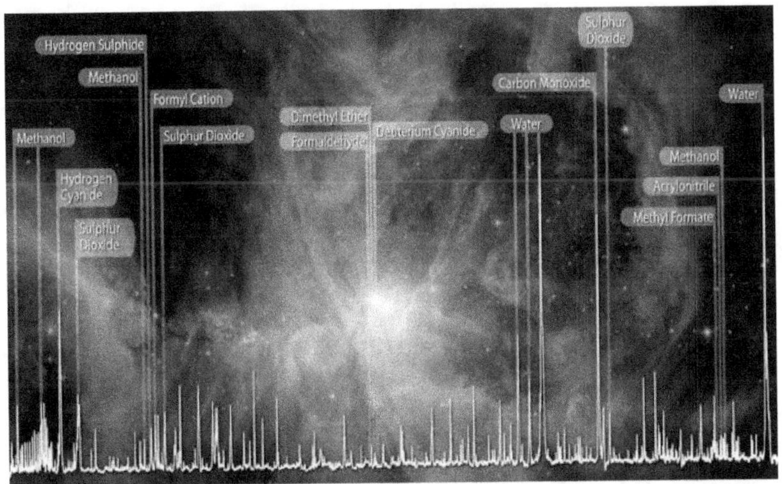

interesting at first, but the information it carries is fascinating. This image shows that the Orion

Nebula is more than just an elemental cloud; intricate chemistry is occurring far out in space.

The peaks on this graph correspond to specific chemical elements, just like the black lines in the Sun's spectrum do. However, some of these peaks are made up of complicated molecules; for example, the nebula contains sulphur dioxide and water. There are further complicated carbon molecules, such as dimethyl ether, formaldehyde, hydrogen cyanide, and methanol, which may come as a surprise. This provides concrete proof that complicated carbon chemistry is taking place in deep space. This is quite intriguing because it suggests that, in a massive cloud of interstellar gas, we are witnessing the chemistry of life as it first forms.

The cosmic connection extends even further; our link to the chemical intricacies of space may be more intimate than we imagine. Pictured here is a meteorite, a fragment of rock that descended to Earth from the vast reaches of the Solar System. With origins likely predating any terrestrial rock, it crystallized from the ancient dust cloud—the nebula—that collapsed eons ago, birthing the Sun and the planets. Upon scrutinizing this time-travelled relic, a revelation emerged: it harbours amino acids, the fundamental building blocks of proteins, which, in turn, constitute the foundation of life.

This discovery hints at the presence of intricate carbon chemistry occurring in the cosmic realms, sculpting the precursors of life over four and a half billion years ago. It tantalizingly proposes that the initial amino acids on Earth might have materialized in the cosmic crucible of space, delivered to our planet via meteorites—an enthralling narrative that blurs the boundaries between the celestial and the terrestrial, showcasing the interconnectedness of the cosmos and life itself.

The Molecular Makeup of Orion: The Herschel Space Observatory of ESA(European Space Agency) captured this comprehensive spectrum, which reveals the intriguing chemical fingerprints of possible organic chemicals that could support life in the Orion Nebula.

This is just one more stunning piece of evidence that compels us to reconsider the gas and dust smudges and glittering lights in the sky. We are gazing at our birthplace as we gaze out into space. We are genuinely descended from the stars, and the entire history of the universe—from the Big

Bang to the present—is encoded in each and every atom and molecule that makes up our bodies.

3. Collapsing

Collapsing, the third chapter of **Marvels of The Cosmos**, explores the fascinating story of gravity, the universe's master sculptor. Although gravity is the weakest of the four fundamental forces in the universe—a mysterious property that beyond our full comprehension—it yet exerts an unmatched amount of power. It has no boundaries; it reaches indefinitely into the wide cosmos and affects everything it comes into contact with. "Collapsing" tells the captivating tale of how gravity, with its mysterious power and pervasive influence, Molds the universe's essential structure and permanently alters the cosmic tapestry.

✛ COMPLETE FORCE

The Universe, with all its vastness and majesty, is governed by the interaction of just four fundamental forces of nature. These forces play a pivotal role in shaping the cosmos as we know it.

Two of these forces, the weak and strong nuclear forces, operate within the atomic nucleus and are not readily apparent in our everyday experiences. They play a crucial role in the stability and behaviour of subatomic particles.

The third force, electromagnetism, is more familiar to us, as it is the force responsible for powering many aspects of our daily lives. Electric currents, for instance, flow due to the influence of electromagnetism. This force governs the behaviour of charged particles and plays a significant role in the interactions between atoms and molecules.

Finally, there is gravity, often referred to as the great sculptor of the Universe. Unlike the other forces, gravity operates on the largest distance scales, shaping the cosmos on a grand scale. From the primordial clouds of hydrogen and helium that pervaded the early universe, gravity orchestrated the formation of the first stars. It played a crucial role in sculpting planets and arranging them into the intricate structures we observe in galaxies.

As gravity continued its influence, it assembled countless solar systems, each with its planets and celestial bodies. Gravity is the driving force behind the intricate dance of moons around planets, planets around stars, and even stars within galaxies. It is the invisible thread that governs the revolutions and orbits of celestial bodies.

On a more personal level, gravity is the force that keeps our feet firmly planted on the ground. It is the reason objects fall when dropped and the force that maintains the balance of the Universe. Gravity's influence extends from the smallest scales, where it governs the behaviour of particles, to the largest scales, where it shapes the structure of the cosmos. In essence, gravity is the silent orchestrator that keeps the Universe ticking over in a harmonious dance of celestial bodies and cosmic phenomena.

For the biggest collections of matter in the universe, the stars, gravity is more than just a benign force; it is unrelenting and has the dual properties of creating and destroying. Gravity crushes the most massive stars out of existence when they run out of nuclear fuel and the other three forces are unable to rearrange the matter in their cores to generate energy and fight its inward pull. Stars shine in temporary resistance to gravitational collapse. It produces the universe's least understood items in the process.

In 1960, Gherman Titov found himself among a select group of twenty men chosen for the Soviet manned space program. His journey through this program involved a rigorous selection process that pushed the limits of both physical and psychological resilience. Among those selected, only two individuals—Yuri Gagarin and Gherman Titov—successfully navigated the demanding challenges of the training.

During their training, Gagarin and Titov, both fighter pilots, demonstrated exceptional skills, matching each other point for point. However, when it came to choosing the first human to venture into space, Yuri Gagarin was given the historic opportunity. On April 12, 1961, Gagarin made history by becoming the first human to travel into space aboard the Vostok 1 spacecraft.

During this groundbreaking mission, Gagarin completed a single orbit around the Earth in a remarkable 108 minutes. Upon re-entry, Gagarin experienced an interesting twist of space trivia: due to concerns about the safety of the capsule during landing, he ejected from the spacecraft at an altitude of 7,000 meters (23,000 feet). As a result, he landed on Earth via parachute approximately 10 minutes after the Vostok 1 capsule had safely touched down.

The significance of Gagarin's achievement cannot be overstated, as it marked a monumental milestone in human space exploration. While Titov did not have the distinction of being the first human in space, the contributions of both Gagarin and Titov were crucial in paving the way for future space endeavours and shaping the course of space exploration history.

Upon his return, Yuri Gagarin achieved iconic status as a Soviet hero and attained global celebrity for being the first human in space. Meanwhile, Gherman Titov, although less well-known, went on to become the second person to orbit the Earth. Titov holds the record as the youngest individual to embark on a space journey, accomplishing this feat at just under 26 years old when he piloted Vostok 2 on August 6, 1961, completing 17 orbits of Earth.

Titov's mission, however, included both historic achievements and some less glamorous moments. During the 25.3-hour mission, he became the first person to sleep in space, taking a nap for a couple of hours as his spacecraft orbited the planet. Additionally, Titov experienced the onset of a condition that would later be termed

Space Adaptation Syndrome, more commonly known as space sickness. This syndrome encompasses various symptoms, including nausea, vomiting, vertigo, and headaches, often arising as a reaction to the unusual sensations experienced during space travel.

Space Adaptation Syndrome has affected nearly half of those who have experienced weightlessness for an extended period. Titov's case highlighted the challenges associated with preparing for weightlessness, which, despite being one of the thrilling aspects of space travel, is difficult to simulate and train for on Earth. Since Titov's mission, space agencies worldwide have used a method similar to what Gagarin and Titov did: creating weightlessness by falling towards Earth. This approach, exemplified by parabolic flights and other simulations, allows astronauts to experience periods of microgravity and aids in preparing them for the unique conditions of space travel.

Project Mercury was the United States' response to the Soviet Vostok program, marking the beginning of American manned space exploration. The project consisted of six manned launches and featured significant milestones, including the historic flight of Alan Shepard on May 5, 1961, making him the first American in space. Another key achievement was the orbital flight of John Glenn, who became the first American to orbit the Earth.

The astronauts chosen for Project Mercury, known as the "Mercury Seven," attained celebrity status in the United States. The members of the Mercury Seven were Alan Shepard, Gus Grissom, John Glenn, Scott Carpenter, Gordon Cooper, Wally Schirra, and Deke Slayton. Each of them eventually flew into space, contributing to the burgeoning space program.

The final mission of the Mercury Seven took place in 1998 when John Glenn, at the age of 77, undertook a Space Shuttle mission. This marked the culmination of the Mercury Seven's remarkable journeys over several decades. The enduring impact of these astronauts extended beyond their space missions, as evidenced by the recognition and admiration they received. In popular culture, the Tracy brothers in the TV series Thunderbirds were named after five of the Mercury Seven astronauts: Scott Carpenter, Gus Grissom,

Alan Shepard, Gordon Cooper, and John Glenn. However, Wally Schirra and Deke Slayton were not included in the namesake selection for the Thunderbirds characters. The era when astronauts enjoyed a level of celebrity akin to rock stars is fondly remembered as a unique and influential period in the history of space exploration.

In response to the experiences reported by Gherman Titov, who was the second human to orbit the Earth and encountered space sickness, NASA implemented a unique training method for would-be astronauts during Project Mercury. This training involved the use of a regular military aircraft, a C-131, to simulate the conditions of weightlessness that astronauts would face in space.

The method employed an unconventional flight path known as a parabolic trajectory. During this flight path, the aircraft would climb steeply, reaching a peak altitude before descending rapidly. This manoeuvre created a brief period, approximately 25 seconds in duration, where all occupants of the plane experienced the sensation of weightlessness. Importantly, during these moments, individuals inside the aircraft were actually in a state of weightlessness.

This parabolic flight path, when repeated numerous times in a single training session, induced physiological effects similar to those encountered in space. Despite the brevity of each weightless episode, the cumulative impact on the human body was significant. To reflect the potential side effects of the experience, particularly the tendency for some individuals to feel nauseous, the C-131 aircraft used for this purpose gained the nickname "Vomit Comet." The term has since become synonymous with any aircraft employed for simulating weightlessness through parabolic flights, and it illustrates the challenging nature of astronaut training, which often requires adapting to unconventional conditions in preparation for space travel.

Since I was a young child and continue to be fascinated about space exploration, I was aware of the Vomit Comet. The Vomit Comet provides a unique environment for experiencing and understanding gravity's fundamental characteristics. It offers insights into two related aspects of the force that are crucial for comprehending its nature.

Firstly, the Vomit Comet demonstrates the ability to nullify the effects of gravity by simulating free fall. Gravity, unlike other forces like electric charge, can be effectively cancelled out by merely falling towards the ground. The aircraft achieves this by following a trajectory analogous to the path a cannonball would take when fired. While the plane doesn't free fall uncontrollably, it is designed to accelerate toward the ground at precisely the same rate as an object in free fall – approximately 9.81 meters per second squared. This acceleration is necessary to counteract the force of gravity and create a weightless environment inside the aircraft. To maintain control, the plane also travels forward at its regular flight speed, resulting in a parabolic flight path.

This phenomenon underscores a fascinating aspect of gravity – in free fall, its effects are entirely eliminated. Whether it's a falling lift or a parachute jump (disregarding air resistance), the sensation of weightlessness is due to the acceleration towards the ground cancelling out the force of gravity. This characteristic sets gravity apart from other forces, as its effects can be selectively removed or enhanced through controlled acceleration.

In the second aspect, the Vomit Comet illustrates that it is also possible to augment the gravitational pull experienced by the occupants. By accelerating the aircraft, additional force is applied, effectively increasing the apparent gravity. This dynamic showcases the two-sided nature of gravity – it can be both neutralized and intensified through controlled motion, providing valuable insights into the force that governs the macroscopic structure of the universe.

Though the reason behind astronauts' weightlessness and ability to float inside their spaceship is not well known, it is a well-known fact. Important point number one: the reason gravity is eliminated by falling is not because they are far from Earth—they are actually only a few hundred miles above the surface, and the strength of Earth's gravitational field in near-Earth orbit is not significantly different from that on the surface.

The experience aboard the modified Boeing 727-200 during weightlessness simulations allowed for a demonstration of a fundamental and intriguing aspect related to gravity, something recognized by Isaac Newton in 1687 and earlier by Galileo. This phenomenon involves the fact that all objects, regardless of their mass, fall at the same rate under the influence of gravity.

Newton's understanding of gravity, expressed in his gravitational theory, highlighted that the gravitational force between two objects is directly proportional to the product of their masses. In simpler terms, the force due to Earth's gravity acting on an object is determined by multiplying the masses of Earth and the object. Consequently, if an object's mass is doubled, the gravitational force it experiences also doubles. However, what seems counterintuitive is that the rate at which an object accelerates toward Earth due to gravity is also directly proportional to its mass. The surprising result is that when the calculations are performed, the mass of the falling object entirely cancels out, leading to the conclusion that all objects fall at the same rate under gravity.

This strange principle was famously demonstrated by Apollo 15 Commander Dave Scott on the Moon's surface in 1971. In the absence of air resistance in the Moon's vacuum, Scott dropped a feather and a hammer simultaneously, and both reached the lunar surface at the same time. On Earth, air resistance causes objects with different masses to fall at different rates, making such a demonstration challenging. The understanding that even a feather and a hammer fall at the same rate under gravity challenges common sense but is a key concept in grasping the nature of gravity.

This principle extends beyond familiar objects to include even beams of light, which, despite being massless, also fall at the same rate as objects with mass under the influence of gravity. This counterintuitive characteristic is a foundational aspect of gravity that, while not aligning with everyday intuition, underscores the universality and fascinating nature of this force.

✛ THE INVISIBLE STRING

Gravity, one of the four fundamental forces of nature, exerts its influence on everything within the cosmos. Its impact ranges from manmade satellites circling Earth, contributing to the technological infrastructure of the twenty-first century, to the natural satellite, the Moon, which completes an orbit around Earth every 27.3 days. Gravity serves as the invisible force that acts like a guiding string, determining the trajectories and paths of celestial bodies.

In our solar system, every planet, moon, rock, and particle of dust is subject to the gravitational pull that shapes their movements. For instance,

Earth's journey around the Sun, completing one orbit every 365 days, is a direct consequence of gravity's influence. The seven planets and 166 known moons in our solar system, each following its unique orbit, are likewise guided by gravity.

Beyond our solar system, gravity continues to play a pivotal role in shaping the dynamics of the Universe. Every cosmic entity, irrespective of its size, experiences the gravitational pull exerted by others. This intricate dance of gravitational forces influences the motion, interactions, and destiny of celestial objects on a grand cosmic scale. Gravity acts as the cosmic orchestrator, conducting the movements and interactions of celestial bodies, from the smallest particles to the most massive galaxies. Its reach is vast, connecting and affecting every corner of the Universe. The profound and universal impact of gravity highlights its role as a fundamental force governing the structure and evolution of the cosmos.

Our solar system revolves around the supermassive black hole at the centre of the Milky Way Galaxy, which is home to nearly 200 billion gravitationally linked stars in a swirling configuration. Furthermore, because gravity continues to guide galaxies around the vast Universe, even this enormous, rotating structure is not the end of the universe's roller coaster.

Our galaxy, the Milky Way, is part of a cosmic congregation known as the Local Group, a cluster of over 30 galaxies first identified by the American astronomer Edwin Hubble in 1936. Spanning over ten million light-years, this extensive dumbbell-shaped structure comprises billions of stars, with the massive Andromeda galaxy being a significant neighbour, boasting trillions of stars.

Within the Local Group, there exists a common centre of gravity around which all member galaxies, including the Milky Way and Andromeda, orbit. This gravitational interplay shapes the intricate dance of these galaxies over vast distances, influencing their positions and movements.

Despite the immense scale of the Local Group, it is not the largest gravitationally bound structure in the cosmos. As you read this, the force of gravity is orchestrating an extraordinary cosmic journey for you. In addition to Earth's daily rotation, orbit around the Sun, and the Sun's orbit within the Milky Way, we are part of a grander gravitational cycle. The entire Milky Way is hurtling around the centre of gravity of the Local Group at a staggering speed of 600 kilometres (372 miles) per second. This dynamic interplay of gravitational forces encompasses various scales, from the celestial dance of galaxies to the intricate movements of stars within them. The intricate gravitational ballet governs our position within the cosmic tapestry, illustrating the profound impact of gravity on the vast structures that define our cosmic neighbourhood.

The Local Group is a member of the Virgo Supercluster, a family of at least 100 galaxy clusters that is significantly larger and gravitationally connected than the Local Group. It is unknown how long it will take our Local Group to travel around the Virgo Supercluster, which is one of the millions of superclusters in the observable Universe. The Virgo Supercluster is enormous, spanning over 110 million light years. These days, it is believed that even superclusters are a component of much larger structures called galaxy filaments or vast walls that are held together by gravity. The Pisces-Cetus Supercluster Complex includes us.

The influence of gravity is infinite and has been present throughout the universe's history at all distance scales. Perhaps surprisingly, though, considering its immense scope and pervasive significance, it is the first force that humans have ever fully comprehended.

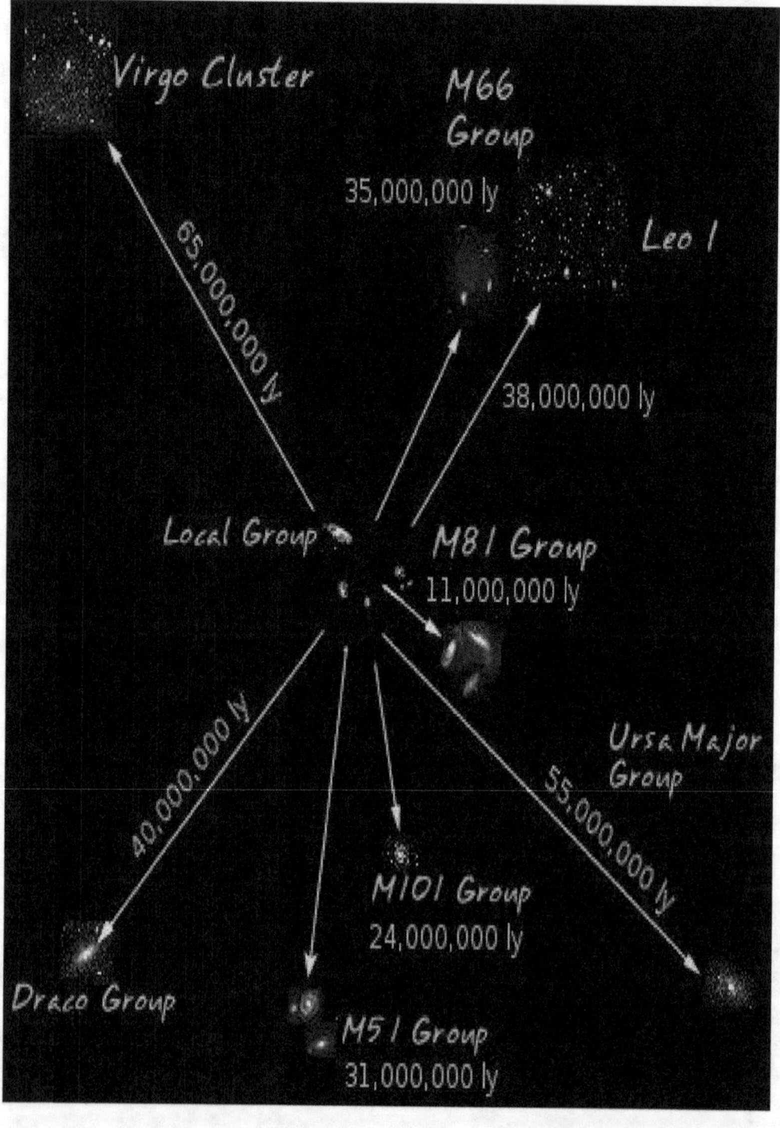

✚ NON-FELLING OBJECT

Curiosity-driven science is the cornerstone of our civilization since the history of science is replete with instances of chance and circumstance leading to the greatest discoveries. The complicated tale of Newton's development of his theory of gravity, the first significant universal law of physics, is one of the most well-known.

The Great Plague of 1665 marked the final significant outbreak of bubonic plague in England and stands out as one of the deadliest episodes of this devastating illness. It is estimated that over one hundred thousand people succumbed to the horrific effects of the rodent-borne disease during this outbreak. The epicentre of the plague was London, but the intricate network of connections between the capital and other regions facilitated the rapid spread of the illness throughout England.

The desperate attempt to contain the plague led to extreme and often futile measures. Various strategies were employed, including the lighting of fires to purify the air and the mass culling of dogs and cats, even though they were not responsible for spreading the disease. Infected villages faced quarantine measures, and preventive actions such as closing schools and colleges were implemented in an effort to curb the spread of the contagion.

Among those affected was Trinity College in Cambridge, where one notable student decided to take a leave of absence during the summer of 1665. That student was none other than Isaac Newton, who, seeking safety from the deadly plague, temporarily withdrew from his academic pursuits. This period of seclusion, however, became a pivotal time for Newton's intellectual development, as he delved into profound scientific inquiries and made groundbreaking discoveries that would later shape the course of physics and mathematics. The Great Plague of 1665 left an indelible mark on history, influencing the lives of countless individuals, including the brilliant mind of Isaac Newton.

At the age of twenty-two, Isaac Newton, having recently graduated, left Cambridge, which was ravaged by the Great Plague of 1665, to return to his family home in Woolsthorpe, Lincolnshire. During this period, he carried with him a collection of books focusing on mathematics, particularly the works of Euclid and Descartes. Newton's interest in these subjects had been sparked by an astronomy book he acquired at a fair. Despite being considered an unremarkable student, his time away from formal education provided him with the space to contemplate and delve into his fascination with the laws governing the physical world.

Over the following two years, Newton's private studies became the foundation for much of his future work. He explored various subjects, including calculus, optics, and, notably, gravity. When he returned to Cambridge in 1667, he was elected as a fellow and eventually assumed the position of Lucanian Professor of Mathematics in October 1670. This prestigious position, previously held by figures like Stephen Hawking and Michael Green, marked the beginning of Newton's influential career in academia.

For the next two decades, Newton engaged in a wide range of scientific pursuits, including both legitimate scientific endeavours and more esoteric interests such as alchemy and predicting the apocalypse. Despite his diverse pursuits, Newton's contributions to modern science are immeasurable. John Maynard Keynes, the economist, described Newton as "the last of the magicians," emphasizing his role on the cusp between pre-scientific times and the modern era.

Newton's most significant contribution to science came in 1687 with the publication of **Philosophiæ Naturalis Principia Mathematica**, commonly known as the Principia. In this groundbreaking work, Newton formulated equations that precisely described the mechanics of gravity. The simplicity and precision of these equations played a crucial role in guiding the Apollo astronauts during their journey to the Moon. Newton's Principia remains a masterpiece of scientific thought, influencing the trajectory of physics and our understanding of the natural world.

Newton's Law of Universal Gravitation, expressed mathematically as:

$$F = G\frac{m_1 m_2}{r^2}$$

This equation states that the force **F** between two objects is equal to the product of their masses m_1 and m_2, divided by the square of the distance **r** between them. The gravitational constant **G** is a proportionality constant that encapsulates the strength of the gravitational force. The gravitational constant is incredibly small (G = 6.67428 times 10^(-11) **N.m²/kg²**), highlighting the weakness of the gravitational force compared to other forces in nature, particularly the electromagnetic force, which is (10^36) times stronger.

The law's beauty lies in its universality, applying everywhere in the universe, except in extreme conditions like the vicinity of black holes or situations involving speeds close to the speed of light. For celestial bodies like planets orbiting stars, stars orbiting galaxies, and galaxies moving in the cosmic expanse, Newton's Law of Universal Gravitation provides accurate predictions. However, in cases where gravity is extremely strong or relativistic effects come into play, Albert Einstein's General Relativity becomes necessary.

One remarkable aspect of this law is its applicability across the entire history of the universe, extending beyond the early moments after the Big Bang. Initially derived from the observations of Tycho Brahe and the laws formulated by Johannes Kepler concerning the motion of planets around the Sun, Newton's law of gravitation unexpectedly turned out to govern the movement of galaxies as well. This universality is a profound statement about the consistency of physical laws throughout the cosmos, and Newton's law of gravitation stands as the first example of such a universal law in the realm of physics.

It's also quite easy. At the core of contemporary fundamental science is the elegant and beautiful idea that the complicated motion of everything in the universe can be encapsulated in a single mathematical formula. It is not necessary to utilize a telescope and trial-and-error methods to determine the locations of the planets and moons in the Solar System each night. Newton's basic equation can be used to determine their location at any given time in the future. This applies not only to our solar system but to all solar systems in the universe. That is the power of physics and mathematics.

From the tiny, immeasurable force of attraction between two rocks on the ground to the much larger force that each of us is currently experiencing between our bodies and the massive rock upon which we are standing, Newton discovered that gravity is a force of attraction that exists between all objects. The force holding each of us to our planet, with our mass of about six billion million million kilogrammes, is sufficient to keep us grounded. However, gravity has far more power on planets than just maintaining orbit and holding objects in place; it may profoundly and unexpectedly sculpt and alter their surfaces.

✛ THE GRAND SCULPTURE

The formation of Fish River Canyon, one of the world's significant geological features, is attributed to the erosive power of water rather than tectonic plate movement or volcanic action. Situated in the south of Namibia, it ranks second only to the Grand Canyon in scale, stretching over 160 kilometres (99 miles) in length, 26 kilometres (16 miles) in width, and reaching depths of up to half a kilometre (a third of a mile) in certain areas.

The key actor in the creation of this impressive canyon is the Fish River, Namibia's longest river, flowing for more than 650 kilometres (403 miles). Despite its seasonal flow limited to the summer months, the river has, over millennia, effectively carved the canyon into solid rock. The process is fundamentally driven by the energy harnessed from the Sun.

The energy transfer begins with the Sun lifting water from the oceans, depositing it upstream in the highlands to the north of the Fish River Canyon. When the rains arrive, gravity takes over. The highlands around the river's source are at an elevation exceeding a thousand meters above sea level. Raindrops, upon landing at this elevation, accumulate gravitational potential energy. The amount of energy stored in each raindrop can be calculated using a simple equation:

U = m.g .h

where:

- **U** is the amount of energy released when the drop falls from a height **h** above sea level to sea level,
- **m** is the mass of the drop, and
- **g** is the acceleration due to gravity (approximately 9.81 m/s^2).

This gravitational potential energy is a crucial driver in the erosive force of water, as it facilitates the carving of the canyon over time. The process is a testament to the dynamic interplay between solar energy, gravity, and the relentless force of water in shaping Earth's geological features.

Because of its location in Earth's gravitational field, every droplet of water hoisted aloft by the Sun carries energy that can be released by letting the water flow downward to the sea. A portion of this energy can be used to chisel away at the surface of the Earth to create the Fish River Canyon.

Earth's gravitational field plays a crucial role in shaping its surface features, influencing not only the behaviour of falling water but also the size of its mountains. Mount Everest, the tallest mountain above sea level on Earth, stands at almost 9 kilometres (5.5 miles), towering above other landforms. However, in the broader context of the Solar System, the most imposing mountain is found on Mars, a planet significantly smaller than Earth.

Mars, located around 78 million kilometres (48 million miles) from Earth, shares similarities with our planet, showcasing evidence of past water activity that sculpted its surface. The Martian landscape bears witness to the historical flow of water from highlands to seas, a process that dissipated gravitational potential energy as the water descended. Despite the absence of liquid water on Mars today, its surface records the geological legacy of previous aqueous interactions.

Mars, with only about 10% of Earth's mass, experiences a significantly weaker gravitational pull. This lower gravity is one of the contributing factors to Mars losing its atmosphere over time, despite its greater distance from the Sun. The reduced gravitational force on Mars led to the gradual escape of its atmosphere, impacting the planet's ability to retain liquid water on its surface. As a consequence, Mars now faces an arid and geologically inactive future.

Interestingly, the lower surface gravity on Mars has a distinctive effect on its mountains. While Earth's mountains reach impressive heights due to the strong gravitational pull, Martian mountains can potentially surpass these elevations. The weakened gravitational force allows for the formation of taller structures, as the balance between the force of gravity and the uplift created by tectonic and volcanic activity results in higher peaks. Thus, the influence of gravity on Martian topography underscores the intricate relationship between a planet's gravitational field and its geological features.

The highest mountain in the Solar System, surpassing every other peak, is Olympus Mons, an extinct volcano located on Mars. Soaring to an impressive altitude of around 24 kilometres (15 miles), Olympus Mons stands almost three times the height of Mount Everest when stacked vertically. This remarkable feature highlights an intriguing aspect of planetary geology: the relationship between a planet's size, gravity, and the height of its mountains.

The extraordinary height of Olympus Mons on Mars is not a mere coincidence. Various environmental factors, including the rate of erosion and the geological history of the planet, contribute to the formation of its distinct topography. However, a fundamental limit to the height of mountains on any celestial body is dictated by the strength of its surface gravity.

Mars, being significantly smaller than Earth with a radius approximately half that of our planet, is also only 10% as massive. Applying Newton's law of universal gravitation, a simple calculation reveals that the gravitational pull at the surface of Mars is approximately 40% of that experienced on Earth. This alteration in gravitational strength has profound implications for the weight of objects on Mars.

The reduced surface gravity on Mars significantly impacts the weight of materials and structures, including mountains. With less gravitational force pulling objects downward, the balance between gravitational forces and the geological processes that create mountains allows for the formation of taller structures. Olympus Mons, reaching an extraordinary height, exemplifies the interplay between a planet's gravity and the geological forces shaping its surface features.

The difference between mass and weight is a fundamental concept in physics, and it becomes particularly significant when considering Einstein's Theory of Special Relativity. Let's break down the explanation:

1. **Mass as an Intrinsic Property**:

  - Mass is an intrinsic property of an object, representing the amount of matter it contains. It is an inherent characteristic and doesn't change regardless of the object's location in the Universe.
  - In everyday terms, the mass of an object is the same whether it is on Earth, the Moon, or in outer space.

2. **Einstein's Theory of Special Relativity**:

  - Einstein's theory introduces the concept of "rest mass," which is the mass of an object when it is at rest (not moving). According to Special Relativity, rest mass is an invariant quantity.

- Invariance means that the rest mass remains the same for an object, irrespective of the observer's location or motion in the Universe.

3. Weight and Gravity:

- Weight, on the other hand, depends on the gravitational force acting on an object. Weight is the force with which an object is pulled towards the centre of a massive body, like a planet.
- On Earth, the weight of an object is proportional to its mass due to gravity. The gravitational force on Earth's surface is approximately 9.81 m/s^2, and the weight (force) is calculated as mass multiplied by the acceleration due to gravity ($W = mg$).

4. Impact of Relativity:

- In regions with significantly different gravitational fields, such as near massive celestial bodies or in strong gravitational fields, the distinction between mass and weight becomes more pronounced.
- Einstein's theory predicts that as an object's velocity approaches the speed of light or as it experiences strong gravitational effects (as near a massive star), its mass can increase,

and the traditional concept of weight becomes less straightforward.

In essence, mass is an intrinsic property that doesn't change, while weight depends on the gravitational force acting on an object. Einstein's Theory of Special Relativity adds depth to these concepts by introducing the idea of invariant rest mass and demonstrating the interplay between mass, energy, and gravitational effects in extreme conditions.

Weight is distinct from mass and is measured in newtons, a unit of force. This distinction becomes evident when considering how weight is measured. When standing on bathroom scales, they gauge the force exerted on them by the individual. This force is directly influenced by the strength of Earth's gravity. The relationship is intuitive – the harder one presses down on the scales, the greater the weight reading.

Symbolically, the weight (W) of an object on Earth is defined as follows: **W = m * g**, where m is the object's mass, and g is the gravitational field strength of Earth, approximately **9.81 m/s^2**. It's essential to note that this definition assumes a constant acceleration due to gravity. To be more precise, weight is defined as the force applied by the scales to provide an acceleration equal to the local acceleration due to gravity, preventing the individual from falling through them.

If, hypothetically, one was to take the scales into a weightless environment, such as the Vomit Comet (an aircraft that creates brief periods of weightlessness), they would register nothing as there would be no force exerted due to gravity. Therefore, weight is contingent on the gravitational field strength of the celestial body. For instance, an individual with a mass of 80 kg on Earth would have a weight of approximately 785 newtons, while on Mars, the same individual would weigh around 295 newtons.

Your weight is influenced by multiple factors, with both your mass and the mass of the celestial body you're on playing crucial roles. Acceleration during weight measurement also contributes, reflecting the equivalence principle. A compelling example is Olympus Mons, the colossal volcano on Mars. If placed on Earth, not only would it surpass all other mountains in height, but its weight would also increase about two and a half times. This immense force would exert such pressure on its base rock that it couldn't support the mountain, causing it to sink into the ground. Earth's gravity prevents it from sustaining a mountain of Olympus Mons' size.

On Earth, Mauna Kea, the dormant volcano in Hawaii, represents the highest point, measured from its base. It is even higher than Everest but is gradually sinking. This sinking phenomenon illustrates the absolute limit imposed by Earth's gravity on mountain height.

The definition of weight becomes intricate due to certain nuances. Earth's gravity varies slightly at different points on its surface, influenced by factors like altitude. Standing on the edge of the Fish River Canyon would result in a slightly lower weight compared to standing at the canyon floor because the gravitational pull is weaker at the top. Earth's uneven density and rotation further complicate matters. Earth's spin introduces acceleration when standing on its surface, adhering to the equivalence principle. This acceleration increases towards the Equator, reducing gravitational acceleration. The bulging at the Equator, caused by Earth's spin, weakens the gravitational pull there even more. Consequently, you weigh approximately 0.5 per cent less at the North and South Poles than at the Equator. The geoid, a precise map, illustrates the effects of Earth's subsurface density variations and surface features on its gravitational field.

The captivating image of Earth's gravitational field was captured by the European Space Agency's satellite, GOCE (Gravity field and steady-state

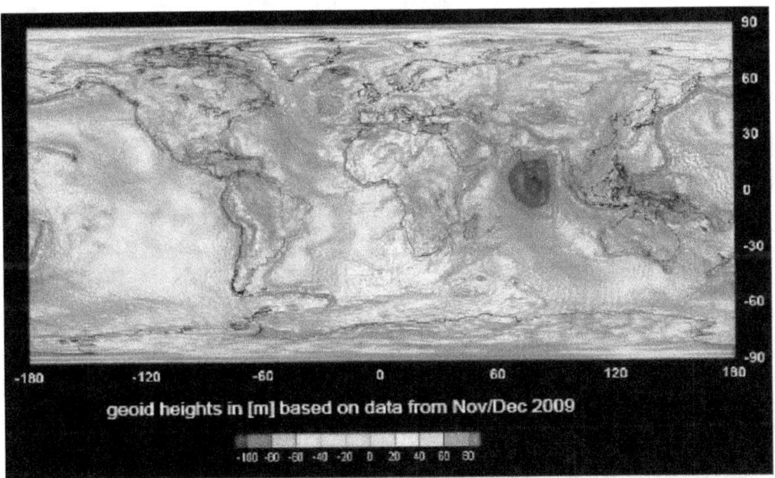

Ocean Circulation Explorer). Launched in March 2009, GOCE is equipped with three ultra-sensitive accelerometers strategically arranged to detect minute changes in Earth's gravitational field as the satellite orbits the planet. Positioned at an altitude of 250 kilometres (155 miles) and skimming the edge of Earth's atmosphere, GOCE spent two months meticulously collecting data to produce this unprecedented gravity map.

This groundbreaking image represents the first-time gravity strength has been mapped globally with such precision. The colour variations in the image convey different levels of gravitational strength: blue patches indicate areas with a weaker gravitational field, green represents average strength, and red signifies places where the gravitational field is stronger. These fluctuations are attributed to variations in the density of rocks beneath Earth's surface and the presence of topographical features such as mountains or ocean trenches.

Technically, the image is presented as an equipotential surface. This means that if Earth were entirely covered by a uniform ocean of water, the image would correspond to the water height at every point. The GOCE satellite's ability to unveil these intricate details in Earth's gravitational field offers valuable insights into the planet's subsurface composition and topographical features, showcasing the powerful impact of gravity on the Earth's structure.

Analysing the gravity map reveals intriguing variations in gravitational field strength, exemplified by the distinction between Iceland and England. Although these discrepancies are imperceptible in our daily experiences, they signify that an individual would weigh slightly less at the same altitude in Manchester compared to Reykjavik.

It's essential to note that the creation of this map was not intended to highlight minor differences in a traveller's weight; instead, its unparalleled level of detail serves as a powerful geological tool with profound implications for understanding Earth's dynamics. The high-precision data derived from the map is particularly beneficial for oceanographers. By defining the baseline water surface in the absence of external factors such as tides, winds, and currents, the map becomes a critical tool for comprehending the intricate factors that influence the movement of water across Earth's oceans.

This detailed understanding of ocean dynamics is crucial for various scientific endeavours. It plays a pivotal role in unravelling the complexities of energy transfer around the planet, contributing significantly to the development of accurate climate models. Ultimately, the gravity map serves as a valuable asset in advancing our understanding of Earth's interconnected systems and enhancing our ability to predict and respond to changes in the global environment.

Therefore, just by measuring the planet's gravity, the geoid provides a wealth of specific information on its structure. However, the most significant aspect of all is not depicted when it comes to the actual height of the ocean surface: the moon.

☩ TUG OF MOON

Families of moons orbit several of the planets in our solar system, including the two small, malformed moons of Mars and the sixty-three satellites of Jupiter, Neptune, and other moons. There is only one moon on our planet, and we have spent nearly four and a half billion years traveling through space with it.

The Moon holds a distinctive status among the celestial bodies in our solar system, with no other planet possessing a moon as substantial in relation to its parent planet as ours. Orbiting at a proximity of merely 380,000 kilometres (236,000 miles) from Earth, the Moon boasts dimensions equivalent to a quarter of Earth's diameter. This renders it the fifth-largest moon in the Solar System, trailing behind Titan, Ganymede, Callisto, and Io—although it's essential to note that the parent planets of these moons, Jupiter and Saturn, significantly surpass Earth in size. This unique dynamic positions Earth and the Moon as a quasi-double-planet system.

The prevailing and widely accepted theory regarding the Moon's formation dates back approximately 4.5 billion years. According to this theory, the Moon originated from the aftermath of a colossal collision between the early Earth and a Mars-sized celestial body known as Theia. This monumental impact resulted in the ejection of debris into orbit around Earth, gradually coalescing into the lunar structure observed today. Evidence supporting this theory is rooted in the Moon's remarkably similar composition to Earth's outer crust. Despite the similarity, the Moon exhibits a lower density, primarily due to its notably smaller iron core. This discrepancy aligns with the notion that the Theia/Earth collision was a glancing blow, preserving Earth's iron core and diminishing the relative iron content in the Moon.

One consequence of the Moon's unique characteristics is its weaker gravitational field compared to Earth. Neil Armstrong's historic step onto the lunar surface during the Apollo mission vividly demonstrated this phenomenon. On the Moon, Armstrong weighed a mere 26 kilograms (58 pounds), a stark contrast to the 81-kilogram (180-pound) weight of his space suit on Earth. This substantial difference in gravitational field strength, approximately one-sixth of Earth's, contributed to the reduced weight experienced by astronauts on the Moon.

Despite its relatively weaker gravitational pull, the Moon exerts a profound influence on Earth. The intricate dance between Earth and its moon results in tidal forces, influencing oceanic movements and contributing to the rhythmic rise and fall of tides. This gravitational interplay has enduring effects on our planet, highlighting the intricate and interconnected nature of celestial bodies within our solar system.

Due to the Moon's proximity to Earth, its gravitational influence varies significantly across the planet. The net gravitational force exerted by the Moon on Earth creates observable effects, particularly in the context of tidal phenomena. The illustration depicts the net gravitational force experienced at each point on Earth as perceived by an observer situated at the Earth's centre, with Earth's own gravitational field subtracted from the calculation.

In the illustration, the gravitational force facing the Moon pulls the side of Earth in that direction toward the Moon, a consequence one would expect. However, there is an equally significant net force acting in the opposite direction, pulling the side of Earth diametrically opposite to the Moon away from it. Additionally, when positioned at right angles to the Moon, the lunar gravity enhances Earth's gravitational pull, causing a compression effect.

This gravitational interplay is the fundamental origin of tides on Earth. As water is more susceptible to stretching than the rigid rock forming the ocean floor, it responds to the varying gravitational forces. Consequently, the water in the oceans bulges outward relative to the ground beneath the Moon and on the opposite side of Earth from the Moon. While the difference in water heights is typically a few meters, it can vary significantly depending on the specific topography of the shoreline.

It's important to note that the influence of lunar gravity extends beyond water bodies, affecting the rocks comprising Earth's surface. Although Earth experiences tidal effects in its rocky structure, the rigid nature of rocks limits their stretching compared to water. As a result, the surface of Earth undergoes minor elevations and depressions, typically on the scale of a few centimetres due to tidal effects.

The continuous rotation of Earth beneath the tidal bulge raised in the oceans leads to the cyclic occurrence of two high tides and two low tides each day. This rhythmic ebb and flow in tidal patterns represent the dynamic interaction between the gravitational forces of the Moon and Earth, showcasing the intricate nature of celestial mechanics.

You will now be justified in wearing an odd, blank, and maybe slightly pitying look on your face the next time someone tries to convince you that since we are formed of water, the Moon must have some sort of effect over us. One is that the tidal effect on you is completely insignificant and has no bearing at all because the difference in the Moon's gravitational force across something the size of your body is negligible. This is because the tides are a differential effect, meaning they depend on the change in the strength of the Moon's gravity across the diameter of Earth. Second, in any event, it has nothing to do with water!

The Earth and Moon have a relationship that is not one-way; just as the Moon's gravity has changed our planet, Earth has also changed its neighbour.

Half of the Moon has been concealed from view throughout human history. The far side of the Moon was last imaged by the Soviet Luna 3 mission in 1959, and this gave us our first look at the otherwise unobserved terrain. Nine years later, the astronauts aboard Apollo 8 made history by being the first people to fly outside of Earth's orbit and to see the far side of the moon with their own eyes. Tidal effects are the reason why only one side of the Moon faces Earth, appearing to be fixed in space and unchanging amid the seemingly endless night sky.

Our spacecraft would have looked substantially different from Earth billions of years ago. When the Moon was still a baby, Earth might have seen both of its surfaces because of its quicker rotation. The Moon was subject to Earth's gravitational pull from the moment of its birth, which would have been stronger given the Moon's closer proximity to the planet at the time.

Gravity is always a two-way street; just as the Moon increases the tides on Earth, the Earth must also cause the tides to sweep across the Moon's surface, according to Newton's Law of Universal Gravitation. Naturally, the solid rock of the lunar surface is the source of these tides rather than water. It would have required incredible planetary heavy lifting to deform the Moon's crust by up to 7 meters (22 feet)!

There was an intriguing impact from this enormous tidal bulge that swept across the Moon. The rock tide was dragged over the Moon's surface as it rotated beneath the massive parent planet suspended in the lunar sky. However, the tide doesn't rise instantly; the Moon's surface needs time to react to Earth's pull. The top of the rock tide will have been carried by the Moon's little rotation at that period. As a result, the tidal bulge will be somewhat ahead of Earth rather than exactly in line with it.

In other words, the Moon's deformity is being pulled back into alignment by Earth's gravity, which functions as a massive break. The phenomenon referred to as tidal locking occurs when the Moon's rotation rate progressively synchronizes with its orbital period. This essentially means that the tidal bulge may stay in one precise location on the Moon's surface underneath Earth and doesn't need to be swept around.

The Moon takes one month to orbit Earth and one month to rotate on its axis because it is currently almost but not quite tidally locked to Earth. Hence, the side of the Moon that is hidden from view receives plenty of sunlight and is always facing away from Earth; it does not have a dark side.

The Moon is actually slowly migrating away from Earth at a rate of less than 4 centimetres (1.5 inches) each year as a result of the Earth-Moon system's ongoing evolution toward perfect tidal locking. This is an intriguing development.

✦ FALSE DAWN

One of the oddest lights to ever shine in our night sky, it has baffled people who have seen its glow for generations, tricking them into believing a new day was approaching. When determining the schedule of daily prayers, the **Prophet Muhammad** cautioned Muslims not to confuse it with the true dawn, referring to it as the **false dawn**.

The magical glow known as the Zodiacal light, which appears on the horizon just before sunrise and after sunset, holds a fascinating connection to the origins of our world and the influential force of gravity. Discovered by the Italian astronomer Giovanni Cassini in 1683, this enigmatic phenomenon has been a subject of intrigue and speculation throughout history.

Initially, scientists were puzzled by the ethereal light, and various theories emerged to explain its origin. One common hypothesis suggested that the light emanated from the Sun's atmosphere as it rose above the horizon before the Sun itself. However, it was Nicolas Fatio de Duillier, a student of Cassini, who provided the groundbreaking explanation for the Zodiacal light.

Fatio de Duillier's insight unveiled the celestial origins of planets and moons within our solar system. The Zodiacal light, he proposed, is sunlight scattered by interplanetary dust particles residing in the plane of the solar system. These fine particles, likely remnants from the formation of planets and other celestial bodies, scatter sunlight in such a way that it creates the characteristic wispy, whitish glow forming a triangular shape on the horizon.

This captivating glow, a ghostly reminder of cosmic history, thus serves as a visual testament to the presence of interplanetary dust and offers a glimpse into the processes that shaped our solar system. The Zodiacal light stands as a poetic interplay of light and matter, connecting us to the distant origins of celestial bodies and the enduring influence of gravity in sculpting the cosmos.

The origin of the Zodiacal light is deeply intertwined with the formative years of our solar system, dating back five billion years to a time when no Sun, planets, or moons existed. Instead, there existed a vast cloud of gas and dust, representing the fundamental building blocks of everything within our solar system today.

The narrative begins with the explosion of a nearby star, sending a shockwave through the cosmic cloud and introducing variations in density and imparting rotation. These density fluctuations played a crucial role in the subsequent development of the solar system. The denser regions, experiencing slightly stronger gravitational pull, initiated a process of growth, with the largest region ultimately giving rise to the Sun.

During its early stages, the solar system lacked the familiar planets and was characterized by a spinning disc of matter surrounding the nascent

Sun—a protoplanetary disc. Over time, the minuscule dust particles within this disc collided and amalgamated, forming larger objects known as planetesimals, akin to small asteroids. The planetesimals, driven by their gravitational forces, began to gather nearby matter, fostering accelerated growth.

Approximately one hundred million years later, the most substantial planetesimals evolved into the diverse array of planets and moons that populate our solar system today. This remarkable journey from a diffuse cloud to a structured planetary system is a testament to the intricate interplay of gravity, collisions, and the dynamic forces that shaped the celestial bodies within the Zodiacal light's captivating glow.

While a significant portion of the matter in the primordial cloud coalesced into planets and moons, not all of it followed this trajectory. Beyond the orbit of Mars, the dynamics of the solar system were set to give rise to another planet. However, a gravitational tug of war ensued between the massive Jupiter and the Sun, preventing the formation of this hypothetical ninth planet. Instead, what remains is a celestial band known as the asteroid belt, composed of countless fragments of rock and debris.

The asteroid belt, ordinarily challenging to observe directly from Earth due to its distance and the diminutive size of its constituents, comes into focus when collisions within the belt generate dust. This seemingly unassuming dust holds the key to the enigmatic glow known as the Zodiacal light. The faint radiance observed after sunset and before sunrise results from sunlight reflecting off the remnants of a thwarted planet—testament to a celestial struggle and a poignant, shimmering testament to the early days of our solar system. The Zodiacal light serves as a celestial canvas, painting a vivid picture of the cosmic forces that shaped our origins.

⊕ BLUE MARBLE

It would be difficult for even the most ardent flat-Earther to explain away **The Blue Marble**. Some have speculated that this stunning photograph of our fragile planet, which was taken by the astronauts on board Apollo 17 during its mission to the Moon on December 7, 1972, may be the most widely shared image in human history. However, why is the Earth a sphere? Why, in fact, are all stars and planets spherical?

The formation of planets and stars arises from the gravitational collapse of vast clouds of dust, and the conceptualization of this process can be approached from different perspectives. While it is common to describe it as the force of gravity pulling matter together, another viewpoint frames it in terms of gravitational potential energy inherent in the particles within the primordial dust cloud.

In essence, each particle in this cosmic ensemble possessed gravitational potential energy as it interacted within the minuscule gravitational fields of neighbouring particles. Analogous to water droplets descending from mountainous heights to minimize gravitational potential energy, these particles sought to descend 'downhill' in the gravitational field to achieve a state of lower potential energy.

This observation leads to a profound and universal principle in physics—a principle that can elegantly explain a multitude of phenomena across the vast expanse of the cosmos. The principle asserts that entities within the Universe tend to minimize their potential energy whenever possible. Whether it's the descent of a ball down a hill or the intricate dance of celestial bodies, this principle provides a foundational understanding: in the grand cosmic ballet, things seek a state of minimum potential energy. Physicists often navigate between the languages of energy and forces, finding interchangeability in expressing the fundamental dynamics of the Universe.

The shape that finally develops in a falling dust cloud will be the one that minimizes the gravitational potential energy. Since anything farther out from the centre of the cloud will have higher gravitational potential energy, the form must allow everything inside the cloud to move as close to the centre as possible! Stars and planets are spherical because the shape that guarantees that everything is as near to the centre as feasible is inherently a sphere.

CODE EXPLORER

The Free Encyclopedia

✛ COLLISION STUDY

Out of the approximately 6,000 stars that are visible to the unaided eye from Earth, just a single object defies the gravitational attraction of our galaxy. The image below shows Andromeda, the most distant object visible to the unaided eye while gazing up at the night sky. It is also the closest spiral galaxy to the Milky Way Galaxy. It may appear as nothing more than a speck in the heavens, but recent investigations by NASA's Spitzer Space Telescope reveal that it is home to a trillion suns.

Andromeda, despite being one among the vast ensemble of a hundred billion galaxies in the observable Universe, stands out for a distinctive trajectory that sets it apart. In contrast to the general cosmic trend where galaxies are propelled away from each other due to the expansion of the Universe, Andromeda is an exception. Rather than participating in this outward cosmic dance, Andromeda is on a direct collision course with our own Milky Way. The two galaxies are gradually converging, and their cosmic rendezvous is propelled by the fundamental force of gravity. This cosmic dance, marked by Andromeda's approach at a rate of approximately half a million kilometres (310,000 miles) per hour, unveils a captivating interplay of celestial forces that will eventually shape the destiny of these galactic neighbours.

Galactic collisions, contrary to the initial perception of rare and catastrophic events, are not uncommon in the vast timescale of the universe. Over billions of years, galaxies, including our neighbours Andromeda and the Milky Way, have undergone mergers and absorbed other galaxies into their structures.

The provided sequence of images offers a computer simulation depicting the hypothetical scenario of a galactic collision between Andromeda and the Milky Way. In this simulation, the Milky Way is shown face-on, progressing from the bottom, moving up to the left of Andromeda, and eventually reaching the upper right. The images span a colossal 1 million light years across, and the timescale between each frame is 90 million years.

Upon the initial collision, both the Milky Way and Andromeda exhibit an open spiral pattern, accompanied by long tidal tails and the formation of a connecting bridge of stars. Following the initial interaction, the galaxies move apart but eventually gravitate back towards each other, culminating in a second collision. The intricate dance of stars results in complex ripple patterns, eventually settling into one massive elliptical galaxy.

While spiral galaxies like Andromeda and the Milky Way represent complexity, order, and beauty, the post-collision state typically yields elliptical galaxies—sterile worlds with few stars forming. The collision between Andromeda and the Milky Way, projected to occur roughly 3 billion years in the future, promises to be a spectacular event. As the galaxies approach, the night sky will be dominated by the presence of our giant neighbour. The colossal energy released during the collision will lead to the formation of vast amounts of stars, illuminating the entire sky in a breathtaking celestial display.

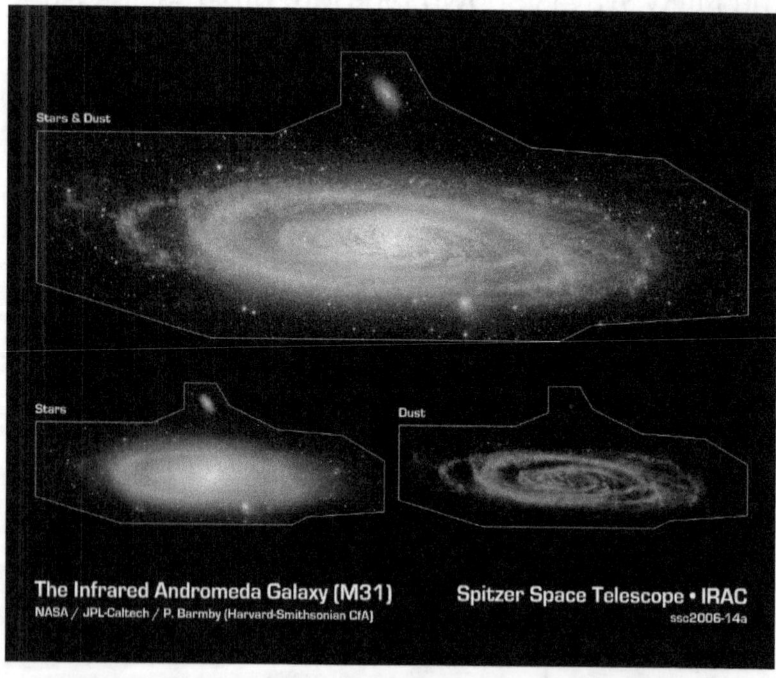

The Infrared Andromeda Galaxy (M31)   Spitzer Space Telescope • IRAC
NASA / JPL-Caltech / P. Barmby (Harvard-Smithsonian CfA)   ssc2006-14a

✛ LAND OF LITTLE GREEN MEN

Using a recently finished radio telescope at Cambridge, postgraduate student Jocelyn Bell and her supervisor Anthony Hewish searched for quasars, the most bright, strong, and energetic objects in the universe, in 1967. It is now commonly accepted that the small, compact regions surrounding supermassive black holes at the centre of extremely young galaxies are what are known as quasars, or quasi-stellar radio emitters. When gas and dust spiral into the black hole, they release an enormous amount of radiation—more than the radiation produced by a galaxy with a trillion suns.

Bell and Hewish discovered an extremely odd signal—a pulse that perfectly reproduced every 1.3373 seconds—while they were searching the data for these extremely active, old galactic centres. The Cambridge team dubbed it LGM-1, or Little Green Men, because they found it nearly difficult to think that such a rapid, consistent pulse could have a natural source.

You would have been notified if they had found evidence of an alien civilization's radio signal. As soon as astronomer Sir Fred Hoyle heard the news, he realized that the source was completely natural. But they had made a breakthrough, and in 1974 Hewish and fellow astronomer Martin Ryle—who, strangely and controversially, was not Bell—were awarded the Nobel Prize in Physics for this finding. It's interesting to note, though, that Bell and Hewish were not the first people to witness one of these marvels; in fact, an ancient civilization that had observed the birth of one almost a thousand years earlier beat them to it.

✦ CHACO CANYON

A millennium ago, spanning the years AD 900 to 1150, an advanced civilization erected a series of immense stone structures known as the Great Houses within the arid Chaco Canyon in New Mexico. These structures, which persisted as the largest manmade constructions in North America until the nineteenth century, showcased remarkable architectural prowess. The largest of these Great Houses boasts over 700 rooms, many of which remain remarkably intact.

Curiously, these structures were not utilized as permanent residences, as evidenced by the absence of traces of fires, cooking tools, or animal bones. Instead, they appear to have served primarily ceremonial purposes. Some archaeologists propose that the intricate architecture of the canyon, including its precisely aligned and complex road system, aimed to symbolize and reinforce the canyon's status as not only the centre of local culture, commerce, and religion but also as the centre of the Universe.

The roads and buildings in the canyon and its surroundings exhibit alignments with the compass points and, speculatively, with significant events in the solar cycle, such as the summer and winter solstices. While the intentional nature of all claimed alignments remains uncertain, it is evident that the Chacon people were astute observers of the skies. They possessed a sophisticated cosmology, complete with intricate stories about the creation of constellations and the Universe itself. The Chaco Canyon civilization left an enduring legacy of architectural marvels and astronomical insights, reflecting their deep connection between the physical world and the celestial realm.

The specific site that drew our attention, concealed about a mile from the primary ruins in Chaco Canyon, held a profound significance for our visit. Ever since my childhood, inspired by Carl

Sagan's Cosmos, I had been aware of and yearned to explore this seemingly unremarkable painting on the underside of a rocky overhang next to a dry riverbed—half a world away from where I first encountered the depiction.

In Sagan's Cosmos, particularly in the chapter titled 'The Lives of the Stars,' there is a small black and white photograph of this painting. The image reveals three symbols: a handprint, a crescent moon, and a bright star. The creation of this painting dates back to approximately AD 1054, coinciding with one of the most remarkable astronomical events documented in history. On July 4, AD 1054, a nearby star underwent a spectacular explosion. Chinese astronomers meticulously recorded the exact date, and it is certain that the Chacoans would have witnessed this celestial phenomenon. The explosion was so intense that it remained visible in daylight for three weeks, and the subsequent new star could be seen with the naked eye at night for two years, dominating the skies.

The precise location of the explosion in the sky is now recognized as one of the most iconic and captivating celestial sights—the Crab Nebula. The painting with the handprint, crescent moon, and bright star stands as a tangible connection to the awe-inspiring cosmic event that captivated the Chacoans over a millennium ago. Whether they celebrated or feared this extraordinary celestial occurrence remains a mystery lost to time. The site holds a unique blend of ancient art and the cosmic spectacle that unfolded in the skies, providing a profound link between the people of Chaco Canyon and the mysteries of the Universe.

The event on July 4, 1054, marked a supernova, the explosive demise of a massive star. Typically, our

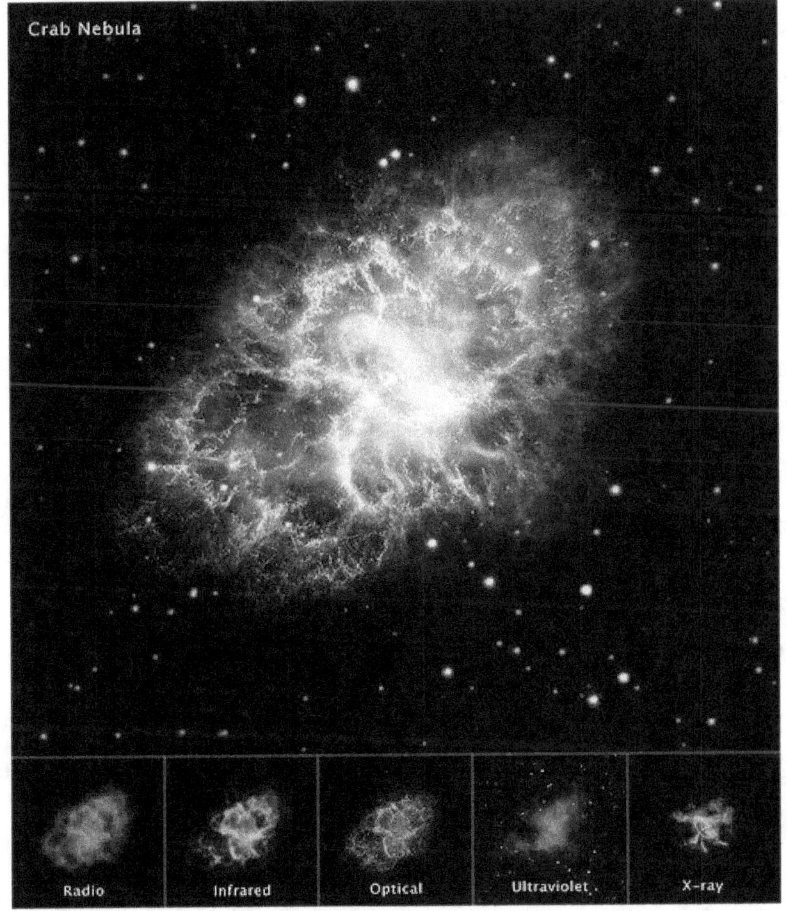

galaxy experiences approximately one supernova per century on average, and the one in 1054 was particularly noteworthy, occurring a mere 6,000 light years away. The aftermath of this cataclysmic event is now manifested in the Crab Nebula—a rapidly expanding structure formed from the remnants of a star that was initially about ten times the mass of our sun.

The Crab Nebula, merely a thousand years after the supernova, spans an impressive 11 light years and is expanding at a rate of 1,500 kilometres per second. At the centre of this luminous cloud resides the exposed stellar core, representing the sole remaining vestige of what was once a massive sun. While the Crab Nebula may appear less remarkable through an optical telescope, the use of a radio telescope unveils a fascinating detail: a pulsing radio signal, precisely occurring 30.2 times per second.

This pulsating signal is attributed to a specific astronomical object—a rapidly rotating neutron star, commonly referred to as a pulsar. The discovery of such pulsars, including the one associated with the Crab Nebula, has been instrumental in advancing our understanding of celestial phenomena. Notably, in 1967, the Cambridge team, led by Jocelyn Bell and her colleagues, detected the distinctive pulsing radio emissions, marking a groundbreaking moment in astrophysics and dispelling any notions of extraterrestrial communication—the team was actually listening to the remarkable sound of a pulsar.

The formation of neutron stars represents a remarkable endpoint in the life cycle of a massive star. Throughout much of a star's existence, the gravitational force pulling matter inward is counteracted by the outward pressure generated from the energy released during nuclear fusion reactions transpiring within its core. This delicate equilibrium sustains the star.

However, when a star exhausts its nuclear fuel, it undergoes a dramatic explosion, shedding its outer layers and leaving behind a core, or stellar remnant. The intriguing question arises: what prevents this remnant from undergoing further gravitational collapse?

The explanation for this phenomenon doesn't reside in the realm of stellar physics but delves into the sub-atomic level of particles. Specifically, neutron stars are formed from the remnants of massive stars. During a supernova explosion, the core of the star undergoes intense compression, and protons and electrons within its atoms merge to form neutrons. This process leads to an extraordinarily dense and compact object, a neutron star.

The strange and fascinating nature of neutron stars arises from their incredible density and the physics governing sub-atomic particles. These remnants serve as a testament to the intricate interplay between gravitational forces and the fundamental particles constituting matter in the cosmos.

The understanding of what prevents normal matter from collapsing inward was clarified in 1967 through the work of physicists Freeman Dyson and Andrew Lenard. They demonstrated that the stability of matter is attributed to a quantum mechanical effect called the Pauli exclusion principle.

In nature, there are two fundamental types of particles distinguished by a property known as spin. The fundamental matter particles, including electrons and quarks, and composite particles such as protons and neutrons, possess half-integer spin and are collectively known as fermions. On the other hand, the fundamental force-carrying particles, like photons, exhibit integer spin and are referred to as bosons.

The Pauli exclusion principle is a crucial property of fermions. It states that no two fermions can occupy the same quantum state simultaneously. In simpler terms, it implies that a large number of fermions cannot be piled into the same location. This principle is fundamental to the stability of atoms and is the underlying reason for chemical reactions.

In atoms, electrons, being fermions, inhabit distinct energy levels or shells around the atomic nucleus. The outermost electrons determine the chemical properties of an element. If it weren't for the Pauli exclusion principle, all electrons would crowd into the lowest energy level, resulting in the absence of complex chemical reactions and, consequently, the absence of the diverse structures and processes that make life possible. The exclusion principle is a fundamental aspect of the quantum nature of matter and has profound implications for the behaviour of particles in the universe.

When attempting to press atoms together, the electron clouds surrounding the atoms are forced into proximity until, at a certain point, the attempt is made to have all the electrons occupy the same quantum state. According to the Pauli exclusion principle, this is prohibited, and it results in an effective force that prevents further compression of the atoms. This force is known as electron degeneracy pressure, and it is a remarkably powerful force in the quantum realm.

In the context of white dwarf stars, which are the remnants of suns that have ceased nuclear fusion in their cores and are gradually cooling, electron degeneracy pressure plays a crucial role. Despite the relentless force of gravity trying to compress the star further, the electrons within the white dwarf resist being forced too closely together due to electron degeneracy pressure. This force counteracts gravitational collapse and maintains a delicate balance, preventing the star from collapsing under its own gravity. The understanding of electron degeneracy pressure is essential in explaining the stability and behaviour of celestial bodies governed by quantum principles.

In the early years of quantum theory, the Indian astrophysicist Subrahmanyan Chandrasekhar conducted a groundbreaking calculation that addressed the fate of massive white dwarfs subjected to increasing gravitational forces. In 1930, Chandrasekhar demonstrated that electron degeneracy pressure, a quantum effect responsible for resisting gravitational collapse in white dwarfs, has its limits.

Chandrasekhar's calculation revealed that electron degeneracy pressure can successfully counteract the collapse of white dwarfs with masses up to 1.38 times that of our sun. However, when attempting to build white dwarfs with masses exceeding this critical limit, the electrons within these stellar remnants reach a point where they can no longer resist the relentless force of gravity. At masses beyond 1.38 solar masses, electron degeneracy pressure proves insufficient, and the electrons ultimately yield to gravity, resulting in their collapse and disappearance. This critical mass limit is known as the Chandrasekhar limit and is a crucial factor in understanding the fate of massive white dwarfs in the cosmos.

When electrons in a collapsing star, particularly in a massive white dwarf exceeding the Chandrasekhar limit, yield to gravity, they don't disappear into nothingness because certain properties, like electric charge, cannot be created or destroyed. Instead, under the intense gravitational force, electrons merge with the protons presents in the atomic nuclei to form neutrons. This transformation is facilitated by the weak nuclear force, operating in a manner opposite to the fusion process in the sun's core where protons convert into neutrons during hydrogen fusion to form helium.

For dying stars surpassing the Chandrasekhar limit, the conversion of the core into a dense sphere of neutrons is the only viable outcome. This compact, neutron-rich structure is known as a neutron star and represents the final state for certain massive stars after they exhaust their nuclear fuel and undergo gravitational collapse.

The matter that constitutes the world around us is predominantly composed of empty space. In a neutron star, the nucleus, which contains nearly all the mass, is incredibly compact. Comparatively, a typical neutron star nucleus is about a hundred thousand times smaller in diameter than its corresponding atom. The majority of the space is occupied by the electron clouds, which are kept at a considerable distance from each other due to the Pauli exclusion principle.

To provide a visual analogy, if the nucleus were the size of a pea, the atom would span a vast sphere approximately a hundred meters in diameter, emphasizing the considerable emptiness within matter. When electrons are removed, gravity causes matter to collapse to the density of the nucleus itself. This gravitational compression eliminates the empty space, resulting in an extraordinarily dense nuclear ball.

A neutron star, typically 1.4 times as massive as the Sun and hovering near the Chandrasekhar limit, is compressed into a remarkably compact sphere, measuring only 20 kilometres (12 miles) in diameter. The density of neutron star matter is so extreme that a mere sugar cube's worth of it would outweigh Mount Everest here on Earth. This highlights the incredible gravitational forces at play within neutron stars, showcasing the remarkable physics governing these celestial objects.

Although much research remains to be done, neutron stars are undoubtedly significantly more complicated than a simple ball of neutrons. The surface gravity is roughly 100,000,000,000G, which is not too much higher than what I felt in the centrifuge. The thin crust of iron and other lighter elements that makes up the surface is most likely composed of iron, but as you go deeper inside, the density of neutrons rises for the reasons mentioned above. It's possible that additional unusual types of matter, such as quark-gluon plasma—an exotic type of pre-nuclear matter that existed in the universe for a few millionths of a second after the Big Bang—exist at temperatures this high deep into the core.

Neutron stars exhibit not only unimaginable density and exotic structures but also possess fascinating features, including intense magnetic fields and rapid rotation. Neutron stars like LGM1 and the one at the centre of the Crab Nebula are known for their powerful magnetic fields. These magnetic fields, resembling those of a bar magnet, are influenced by the stars' rotation. If the magnetic axis is tilted relative to the spin axis, it gives rise to two high-energy beams of radiation that sweep around akin to lighthouse beams.

This phenomenon results in the creation of pulsars, a term coined for stars exhibiting these pulsating beams of radiation. The intricate details of this mechanism are the focus of intensive theoretical and experimental research. The stars observed by Bell and Hewitt in 1967 are examples of pulsars. The fastest known pulsars, called millisecond pulsars, astonishingly rotate over a thousand times every second. This rapid rotation, occurring in a star compacted to the size of a city and resembling a single atomic nucleus, underscores the extraordinary violence and extreme conditions present in these celestial objects.

In January 2004, a groundbreaking discovery was announced by astronomers utilizing the Lovell Telescope at the Jodrell Bank Centre for Astrophysics near Manchester and the Parkes Radio Telescope in Australia. They identified a double pulsar system, a phenomenon considered one of the most extraordinary wonders in the Universe. This system comprises two pulsars, one with a rotational period of 23 thousandths of a second and the other with a period of 2.8 seconds. The two pulsars orbit each other every 2.4 hours, and the diameter of their orbit is so compact that the entire system could comfortably fit inside our sun.

Pulsars, being incredibly accurate timekeepers, provided astronomers with a unique opportunity to test Einstein's theory of gravity under the most extreme conditions known. The double pulsar system, with its massive, spinning neutron stars, allowed scientists to investigate the intense warping and bending of space and time in proximity to these celestial objects. In a remarkable testament to the power of human intellect, Einstein's Theory of General Relativity, formulated by a man living at the turn of the twentieth century, has been confirmed with exceptional precision in the double pulsar system—accurate to better than 0.05 percent. This achievement showcases the majestic, powerful, and wonderful capabilities of the human intellect, demonstrating how a theory inspired by contemplating falling rocks and elevators can accurately describe the motion of the most exotic objects in the Universe, even in the most extreme conditions. This profound connection to the fundamental principles of the cosmos is one of the reasons for the deep admiration and love for physics.

⚓ WHAT GRAVITY IS?

Newton revolutionized our knowledge of the universe when he published his Law of Universal Gravitation for the first time in 1687. As we've seen, his straightforward mathematical formula can accurately predict how moons orbit planets, planets orbit the Sun, solar systems orbit galaxies, and galaxies revolve around one another. Newton's law, however, is merely a model of gravity; it says nothing about the nature of gravity itself, and it surely says nothing about a fundamental question: why do all objects fall in gravitational fields at the same speed? We can rephrase this question by taking another look at Newton's well-known equation: $F = G.m_1.m_2/r^2$

Newton's law of universal gravitation is a fundamental principle in physics, describing the gravitational force between two objects as proportional to the product of their masses. In mathematical terms, if we denote the mass of Earth as m_1 and the mass of an object falling towards Earth as m_2, the gravitational force F is given by $F = G.m_1.m_2/r^2$, where G is the gravitational constant and r is the separation between the centres of the two masses. However, when we delve into Newton's Second Law of Motion $F = ma$, which expresses the relationship between force **F**, mass **m**, and acceleration **a**, we encounter an interesting connection.

Newton's Second Law can be rearranged to **a = F/m**, indicating that the acceleration of an object is determined by the force applied to it divided by its mass. Notably, when we consider an object falling under gravity, the acceleration **a** becomes equal to the gravitational force F acting on the object divided by its mass m. What's intriguing is that the mass term in both equations m_2 in the gravitational force equation and **m** in the acceleration equation) cancels out, resulting in an acceleration that depends only on the mass of Earth. This observation leads to the well-known gravitational acceleration of **9.81m/s^2**regardless of the object's mass.

However, a critical question arises: why should the inertial mass (related to how difficult it is to accelerate an object) in the acceleration equation have any correlation with the gravitational mass (related to how gravity acts on an object) in the gravitational force equation? This assumption lacks a fundamental justification in Newtonian physics and raises intriguing questions about the nature of mass and its connection to gravity. Newton, despite his groundbreaking contributions, did not provide a satisfactory answer to this pivotal question.

Sir Isaac Newton's laws of motion and the law of universal gravitation, formulated in the 17th century, provided an incredibly successful model for understanding the motion of objects under the influence of gravity. Newton's laws became the cornerstone of classical physics, accurately predicting the behaviour of celestial bodies and everyday objects alike. However, Newton refrained from delving into the underlying nature of gravity, describing it as the work of God. Despite lacking a clear explanation of what gravity truly is, Newton's model proved remarkably effective in explaining and predicting a wide array of observational phenomena.

The enigma of gravity's true essence caught the attention of Albert Einstein, who, unsatisfied with the lack of a fundamental explanation, embarked on a quest for a more profound understanding. Building on the groundwork laid by his Special Theory of Relativity in 1905, which revolutionized our understanding of space and time, Einstein sought to extend his revolutionary insights to the realm of gravity.

One of Einstein's key considerations was the equivalence between gravitational and inertial mass, a concept he explored deeply. This equivalence suggested a fundamental connection between the force of gravity and the effects of acceleration. This line of thinking led Einstein to develop the General Theory of Relativity, presented in 1915. This groundbreaking theory redefined gravity as the curvature of spacetime caused by the mass and energy present in it. Rather than a force acting at a distance, gravity became a consequence of the geometry of the fabric of spacetime itself.

Einstein's General Theory of Relativity not only provided a more comprehensive explanation of gravity but also offered a profound shift in our understanding of the universe. It successfully accounted for phenomena unexplained by Newtonian gravity, such as the bending of light around massive objects (gravitational lensing) and the subtle shifts in the orbits of planets. The theory has stood the test of time, with its predictions verified through numerous experiments and observations.

In summary, while Newton's model served as a brilliant mathematical framework for gravitational interactions, it was Einstein who, through the

General Theory of Relativity, offered a deeper and more profound conceptualization of gravity, reshaping our understanding of the fundamental forces governing the cosmos.

Even if he wasn't particularly driven by it, Einstein would have realized that Newton's theory had issues that went beyond philosophical ones. The most unnerving of them was the clearly troubling behaviour of a rock ball that was more than 48 million miles (77 million kilometres) away from Earth.

Mercury, the closest planet to the Sun, has captivated astronomers for centuries due to its unique characteristics and challenging observational conditions. Its orbit, characterized by extreme temperature variations, eccentricity, and proximity to the Sun, makes it a subject of fascination and difficulty in observation.

The planet's orbit is highly eccentric, meaning it deviates significantly from a perfect circle. At its closest approach to the Sun, Mercury is a mere 46 million kilometres (28 million miles) away, while at its farthest, it retreats to over 69 million kilometres (42 million miles). This elliptical orbit results in variable speeds during its journey around the Sun, demanding high-precision measurements to accurately map its path and predict future astronomical events like transits.

One of the most captivating phenomena involving Mercury is its occasional transit across the face of the Sun as observed from Earth. These transits, occurring only 13 or 14 times every century, are celestial spectacles that have drawn the attention of scientists throughout history. During the seventeenth and eighteenth centuries, astronomers gathered globally to witness these rare events. Armed with Sir Isaac Newton's Law of Gravity, they made predictions about the timing and locations of Mercury's transits. However, these predictions sometimes faced challenges when the planet seemed to defy expectations, crossing the Sun's disc later than anticipated, occasionally by several hours.

Despite the challenges posed by its elusive nature, the study of Mercury, its transits, and the precision required for accurate predictions have contributed significantly to our understanding of celestial mechanics. These observations serve as reminders of the intricate dance of the planets in our solar system and the ongoing efforts to unravel their mysteries.

Mercury's peculiar orbit presented a significant challenge to astronomers, and it wasn't until 1859 that Urbain Le Verrier, a French astronomer, demonstrated that the intricacies of Mercury's orbit could not be entirely accounted for by Newtonian gravity. This realization marked a departure from the previously held belief that Newton's laws of gravity could precisely explain the motion of all celestial bodies.

In an attempt to resolve the discrepancies in Mercury's orbit, astronomers hypothesized the existence of another unseen planet orbiting between Mercury and the Sun. Despite being invisible to telescopes of that era, this hypothetical planet, dubbed Vulcan, was presumed to exert a gravitational influence significant enough to perturb Mercury's orbit.

The inspiration for this search came from the successful prediction and subsequent discovery of Neptune, which had been identified based on anomalies in the orbit of Uranus. Encouraged by this precedent, astronomers embarked on a quest to find Vulcan, anticipating that its presence would provide a solution to the observed irregularities in Mercury's trajectory.

The pursuit of Vulcan, however, faced challenges, and despite numerous efforts and claims of sightings, the elusive planet remained elusive. The eventual resolution to the mysteries of Mercury's orbit would come later with the advent of Albert Einstein's theory of General Relativity, which offered a more accurate and comprehensive explanation for the gravitational interactions in the solar system, including the peculiarities of Mercury's movement.

⊥ THEORY OF GENERAL RELATIVITY

Einstein's fascination with the concept of freefall, a fundamental aspect of his thought experiments leading to General Relativity, would have found a unique expression in the experience of the Vomit Comet. The profound realization that gravity's effects can be completely nullified by freefall within a gravitational field, a key insight for Einstein, takes on a tangible and immersive quality in the weightless environment created by the parabolic flight of the Vomit Comet.

As I floated alongside a small plastic Albert in the Vomit Comet, the essence of Einstein's curiosity about freefall became vividly apparent. Within the descending plane, approaching Earth, the sensation is utterly deceptive. There's an impossibility in discerning the motion, proximity to a planet, or the acceleration of 9.81 m/s^2 toward the ground, as perceived by someone on solid ground. Everything inside the plane, including myself, water droplets released from a bottle, the cameraman, director, and the miniature plastic Albert, shared a state of weightlessness. The absence of discernible forces was self-evident, as nothing exhibited movement that would indicate the presence of an external force.

This experiential understanding of freefall within the Vomit Comet encapsulates the intrigue that captivated Einstein, further emphasizing the profound implications of his thoughts on gravity and its interplay with the dynamics of falling objects.

The paradoxical nature of the Vomit Comet experience, seen both from inside and outside, resonates deeply with Einstein's theory of General Relativity, particularly the equivalence principle. While floating weightlessly inside the plane, the occupants, including myself, are oblivious to the effects of gravity and feel no acceleration towards the ground. Yet, viewed from the ground, the plane is engaged in a high-speed, accelerating descent with the unmistakable force of gravity at play.

Einstein's theory asserts that these two perspectives, the internal and external views of the Vomit Comet's motion, are equally valid. The individuals inside the plane are justified in claiming there's no experiment proving their acceleration toward the ground, effectively cancelling out the force of gravity. Even gazing out of the windows doesn't provide a conclusive counterargument, as one could propose an alternative scenario where Earth is accelerating upward, rendering the occupants in a state of apparent weightlessness.

This fundamental insight aligns with the equivalence principle, a cornerstone of Einstein's General Relativity, emphasizing the indistinguishability between the effects of acceleration and gravity. Essentially, the feeling of gravitational force experienced by everyone on Earth can be equated to the upward acceleration of 9.81 m/s^2, illustrating the profound connection between acceleration and gravity, a concept central to Einstein's revolutionary theory.

In technical terms, Einstein would have described the Vomit Comet as an inertial frame of reference during its freefall, meaning that it is genuinely at rest and not subject to external influences.

All objects fall at the same rate, which can be explained by the claim that sitting in a falling airplane is absolutely comparable to drifting around in space, far beyond the gravitational attraction of any planet or moon.

T

he perplexing yet intriguing scenario within the Vomit Comet, with its dual perspectives inside and outside the plane, encapsulates the essence of Einstein's profound insights into gravity and acceleration. From within the plane, the occupants experience a weightless, seemingly force-free environment, giving the impression that nothing is happening—no forces are at play. This tranquillity is disrupted when viewed from the external vantage point, revealing a swift descent towards Earth under the influence of gravity.

The crux lies in the fact that reality remains consistent regardless of the observer's viewpoint. Whether inside or outside the plane, the behaviour of objects, such as plastic Albert and water droplets, must conform to the same laws of physics. If they float weightlessly inside the plane from one perspective, they should exhibit the same weightlessness when viewed from an external standpoint.

Crucially, this analysis doesn't rely on assumptions about the equivalence between gravitational and inertial masses. It simply establishes that a freely falling object within Earth's gravitational field is indistinguishable from one in freefall in space or anywhere else in the Universe, irrespective of celestial bodies nearby.

Einstein's elegant explanation in General Relativity adds a new layer of understanding to gravity. Gravity, according to his theory, is not a mysterious force acting at a distance but rather the result of the curvature of spacetime—the very fabric of the Universe itself. This conceptual shift transforms our perception of gravity from a force between masses to a manifestation of the geometry of the spacetime continuum.

To comprehend the curvature of spacetime, a helpful analogy can be drawn from the familiar curved surface of Earth. While Earth's surface is inherently three-dimensional, let's simplify our perspective to consider it as a two-dimensional space with coordinates defined by latitude and longitude.

In our three-dimensional existence, we can easily perceive Earth's curvature as a sphere. However, imagine a hypothetical scenario where we are two-dimensional beings, confined to navigating only on the surface of Earth, with no awareness of a third dimension—no notion of up or down. In this restricted perspective, our understanding would be limited to movements along latitude and longitude lines.

As two-dimensional beings, visualizing the curvature of Earth's surface becomes challenging. The concept of the planet's three-dimensional curvature would be elusive, as our perceptual constraints prevent us from grasping the full depth of the spatial arrangement. Much like these hypothetical beings, our understanding of the curvature of spacetime is influenced by our inherent limitations in visualizing dimensions beyond our three-dimensional experience.

This analogy lays the groundwork for comprehending the intricacies of spacetime curvature, where our usual spatial intuitions are challenged by the peculiar nature of the cosmos. Just as the two-dimensional beings struggle to envision the curvature of Earth's surface, our attempts to visualize the curvature of spacetime encounter similar cognitive challenges due to the abstract nature of higher-dimensional geometries.

Expanding our analogy, let's delve into how the curvature of a surface can give rise to a force, drawing inspiration from a pair of two-dimensional friends on Earth's Equator. These friends decide to embark on a journey due north while pledging to walk parallel to each other, ensuring they never collide. Their plan is to follow the lines of longitude as they head towards the North Pole.

In their two-dimensional existence, these friends perceive themselves as moving along straight, parallel paths. However, as they progress northward and approach the North Pole, something peculiar happens—they find themselves getting closer and closer to each other. The ultimate surprise occurs when, upon reaching the North Pole, they unexpectedly collide. The conundrum perplexes them, as they meticulously adhered to their parallel trajectories.

From the vantage point of three-dimensional beings, we can elucidate this phenomenon—the surface of Earth is curved. In the three-dimensional space we inhabit, all lines of longitude converge at the poles due to the spherical curvature of Earth. However, for our two-dimensional friends, unaware of the third dimension and the curvature of their surface, this peculiar convergence gives the illusion of an inexplicable force drawing them together.

In Einstein's theory of General Relativity, this force is identified as gravity. Analogously, just as our two-dimensional friends felt a mysterious force compelling them to converge, objects in our three-dimensional reality, including planets and stars, experience gravitational forces due to the curvature of spacetime caused by the presence of massive objects. This conceptual leap from the observable effects on Earth's surface to the broader cosmic context exemplifies the intricate relationship between curvature and gravitational forces in Einstein's revolutionary theory.

Einstein's Theory of General Relativity introduces a complex concept—the four-dimensional surface known as spacetime, comprising the three familiar dimensions of space intertwined with an additional dimension of time. Though delving into the intricacies of spacetime exceeds our current scope, it's crucial to acknowledge its necessity in explaining various phenomena, particularly the behaviour of light and the structure of Maxwell's equations discussed earlier.

In essence, our universe's surface, where our daily lives unfold, is characterized by four dimensions according to Einstein's formulation. What Einstein illuminated is that the presence of matter and energy—manifested as stars, planets, and moons—exerts a curvature on the spacetime surface, sculpting it into hills and valleys. His equations precisely define the shape of spacetime around specific objects, such as the Sun, and dictate the trajectories of objects navigating this curved surface.

The pivotal realization is that, akin to the movement of our two-dimensional friends on Earth's surface, objects traverse spacetime in straight lines. However, this straightforward interpretation is deceptive when one overlooks the curvature of spacetime. In the context of this curved surface, the seemingly straight paths of objects create an illusion of external forces at play, distorting their trajectories.

Applying his geometric theory of gravity, Einstein conducted a groundbreaking calculation concerning Mercury's trajectory through the curved spacetime around the Sun. His findings aligned perfectly with centuries of observations during transits, showcasing that Mercury would orbit the Sun in a manner consistent with empirical evidence. This triumph marked a stark departure from Newton's gravitational framework, as Einstein's theory provided a more accurate description of celestial motions, particularly in scenarios where gravitational forces were notably strong, as in the vicinity of the Sun.

Einstein discovered a very beautiful and entirely geometric method of explaining the force of gravity. It not only forecasts Mercury's orbit, but it also offers a compelling justification for the equivalency principle. In a gravitational field, why do all objects descend at the same speed regardless of their mass or composition? Because they are only following straight lines through the curved spacetime, the path they take has nothing to do with them.

The most astonishing illustration of Einstein's theory of General Relativity lies in the phenomenon of light bending due to gravity. This concept challenges Newtonian physics, as light, lacking mass, should ostensibly remain unaffected by gravity within the Newtonian framework. However, Einstein's theory introduces a profound shift in perspective. According to General Relativity, light, despite its lack of mass, follows a straight path through the curved spacetime influenced by gravity. Consequently, light appears to traverse the same trajectory as any other object subject to gravitational forces.

To comprehend the peculiarity of this idea, consider a thought experiment conducted on a colossal planet. Stand on its surface with a rock in one hand and a laser beam in the other. Orient the laser beam horizontally, release the rock, and emit the laser simultaneously. Counterintuitively, both the rock and the laser beam hit the ground simultaneously, as they navigate through the same curved spacetime. This reveals a fascinating equivalence: in a gravitational field, light descends at the same rate as other objects.

However, there is a crucial caveat. The reference to a "very, very big planet" is deliberate, considering that light travels nearly 300,000 kilometres per second. In the scenario where the rock takes a second to reach the ground, the light would have covered an extensive horizontal distance. On Earth, this would result in the surface curving away from the light's path. Nevertheless, the fundamental principle remains intact, emphasizing the parity in the gravitational descent of light and material objects.

Firing a laser beam directly at the ground introduces a fascinating aspect of General Relativity, particularly in the context of light's behaviour. Despite the constant speed of light—299,792,458 meters per second—it encounters a peculiar situation when subjected to the influence of gravity. In the scenario of aiming the laser beam at the ground, where it should theoretically experience acceleration due to gravity (9.81 m/s^2), a paradox arises. Light cannot alter its speed, but its energy can undergo changes.

As the light descends toward the ground, it gains energy from its fall, leading to a shift towards the blue end of the spectrum. This spectral shift signifies a reduction in wavelength and an increase in frequency. This observation holds significant implications, especially in the context of time measurement. If the frequency of the laser beam synchronizes a clock, firing the laser at the ground results in an increased frequency upon impact. Consequently, the clock located above the ground, which relies on this synchronized laser frequency, appears to run slightly fast from the perspective of an observer on the ground.

This intriguing phenomenon is known as gravitational time dilation—an effect where gravity influences the passage of time. In the framework of General Relativity, it can be expressed as the bending of spacetime near massive objects, such as Earth, causing time to pass more slowly in their vicinity compared to regions farther away. The practical application of this concept is evident in the GPS satellite navigation system. GPS satellites, orbiting at an altitude of 20,000 kilometres, experience a weaker gravitational field than the Earth's surface. As a result, their clocks run faster by 45 microseconds per day, accounting for both gravitational and relativistic effects. This meticulous consideration of time variations is essential to maintaining the accuracy of the GPS system, illustrating the real-world relevance of Einstein's gravitational theories in our daily lives.

In conclusion, if Einstein had seen the Vomit Comet, he would have said that it fell through spacetime in a straight line at that period. The aircraft and its occupants will not experience any force of gravity as long as it stays on this course; a force is only experienced when anything causes the plane to deviate from its journey through spacetime. The ground would be this impediment if the plane didn't stop falling!

The equivalence of gravitational and inertial masses, as experimentally observed, has been a cornerstone of the discussion on the nature of gravity and motion. Einstein's General Relativity offers a compelling explanation for this equivalence by introducing the concept of spacetime curvature. According to Einstein, gravity is not a mysterious force acting at a distance but rather a consequence of the curvature of spacetime caused by the presence of matter and energy.

In this geometric interpretation, objects move along straight lines in the curved spacetime, and the apparent gravitational force arises from their deviation from what would be considered a straight path in flat space. This elegant framework provides a unified understanding of gravity as an intrinsic feature of the geometry of the universe.

However, it is essential to acknowledge that alternative perspectives exist. One such viewpoint suggests that the equality of gravitational and inertial masses may have a deeper, yet undiscovered, underlying reason. In this scenario, the equivalence is seen as a fundamental aspect of the relationship between gravity and inertia.As of now, there is no definitive way to choose between these interpretations, and both are considered valid ways of approaching the understanding of gravity. Whether one sees General Relativity as a geometric theory or a model akin to Newton's theory depends on the philosophical stance taken regarding the fundamental nature of physical laws. This openness to different perspectives highlights the ongoing exploration and complexity of our understanding of gravity and the fabric of spacetime.

Indeed, the beauty of Einstein's General Relativity lies not only in its conceptual elegance but also in its ability to make predictions that align with observations of the natural world. While phenomena like the orbit of Mercury and gravitational time dilation near massive bodies provide compelling evidence for the theory, testing it under extreme conditions becomes crucial to truly pushing its limits.

To subject General Relativity to the most rigorous tests, scientists turn their attention to the most exotic and massive objects in the Universe. These objects, often found in distant reaches of space, exert gravitational forces of exceptional strength. By studying the behaviour of light, matter, and spacetime in the intense gravitational fields near these objects, researchers aim to verify whether the predictions of General Relativity hold up even in the most extreme scenarios.This pursuit takes astronomers and physicists to places like neutron stars and black holes—entities with gravitational fields so intense that they significantly warp the fabric of spacetime. Observations of light bending around these massive objects, gravitational lensing effects, and the behaviour of matter in their vicinity become crucial tests for the accuracy of Einstein's theory under extreme gravitational conditions.

The quest to understand the fundamental nature of gravity and spacetime continues to drive scientific exploration and inquiry, taking us to the boundaries of our current understanding and challenging the very fabric of the universe as described by General Relativity.

✦ INTO THE DARKNESS

One of the greatest human achievements, in my opinion, is the success of Einstein's Theory of General Relativity, and it will remain such for as long as there is anything that can be referred to be a civilization. The tale of gravity, however, has one last twist: Einstein's groundbreaking theory foretells its own destruction.

The collapse of a neutron star is a fascinating process governed by the delicate balance between gravitational forces and quantum mechanical principles. Neutron degeneracy pressure plays a crucial role in preventing the complete collapse of a neutron star. Neutrons, being fermions like electrons, possess a property known as spin. The Pauli exclusion principle dictates that no two fermions can occupy the same quantum state, preventing them from being squeezed too closely together.

In the case of a neutron star, the immense gravitational forces resulting from the star's mass are counteracted by neutron degeneracy pressure. Neutrons, being more massive than electrons, can be densely packed before the Pauli exclusion principle comes into play, resisting further contraction. This creates a stable state where the star maintains a delicate equilibrium, resisting gravitational collapse up to a certain point.

However, the story doesn't end with neutron degeneracy pressure. Theoretically, quark degeneracy pressure could serve as another layer of stability against gravity. Quarks, like neutrons and electrons, are fermions, and their properties contribute to the overall resistance against gravitational collapse. Quark matter is hypothesized to exist at extreme densities and pressures, potentially forming a state of matter known as quark-gluon plasma.

Despite these remarkable mechanisms, there is a limit beyond which known physical laws cannot counteract gravity's relentless pull. The Tolman-Oppenheimer-Volkoff (TOV) limit, estimated to be around three times the mass of the Sun, represents this critical threshold. Beyond this limit, even the combined resistance of neutron and quark degeneracy pressure is insufficient to prevent gravitational collapse. Stars exceeding this mass, if they undergo supernova events, may lead to the formation of black holes—a domain where gravitational forces become so dominant that not even light can escape.

Understanding these processes sheds light on the fascinating interplay between quantum mechanics and gravity in extreme astrophysical environments. The exploration of neutron stars and the theoretical limits of degeneracy pressure contribute to our broader comprehension of the life cycles and fates of massive stellar remnants in the cosmos.

In the early days of General Relativity, Karl Schwarzschild made a groundbreaking contribution just one month after Einstein published the theory in 1915. He discovered a solution to Einstein's equations, now known as the Schwarzschild metric, which describes the structure of spacetime around a spherically symmetric object. This metric has two noteworthy features, particularly when considering the region around a massive object.

The first significant feature is associated with a specific distance from the massive object, known as the Schwarzschild radius. For distances less than the Schwarzschild radius, spacetime becomes distorted in a peculiar way: the entire future of anything that falls beyond this radius is directed inwards. In the context of Einstein's theory, where space and time are intertwined, this distortion implies that the future light cones inside the Schwarzschild radius point inexorably towards the centre. This peculiar phenomenon means that once an object or even light crosses this boundary, it is destined to move inward towards the massive object, with no possibility of escape. This defining boundary is termed the event horizon.

The second intriguing aspect of the Schwarzschild metric involves considering the fate of the massive object itself. Taking the Sun as an example, if one were to calculate the Schwarzschild radius for a star with the mass of the Sun, it turns out to be approximately 3 kilometres (1 mile). This raises an interesting point because, theoretically, this radius is located inside the Sun. However, this is not problematic since, in reality, you cannot get that close to the Sun without already being within its outer layers. Once inside the Sun, the mass outside you becomes irrelevant in the gravitational context.

These insights into the Schwarzschild metric offer a glimpse into the strange and captivating nature of spacetime near massive objects, providing a foundation for understanding phenomena like event horizons and the gravitational behaviour of extremely dense celestial bodies.

When considering an object like a collapsing neutron star, undergoing a relentless process of shrinking and increasing in density, a fascinating question arises: what if this object became so dense that it could possess the mass of the Sun while being physically smaller than the Schwarzschild radius? The answer to this intriguing scenario lies in the concept of black holes, a phenomenon believed to exist in the vast expanses of the Universe.

Black holes represent extreme gravitational environments where even the resistance provided by neutron degeneracy pressure is insufficient to counteract the relentless pull of gravity. These celestial entities are characterized by their ability to compress immense mass into an exceptionally small volume, potentially smaller than the Schwarzschild radius for that particular mass.

Delving into the heart of a black hole, at the point $r=0$ in the Schwarzschild metric, a profound revelation emerges. The curvature of spacetime reaches infinity, indicating an infinite gravitational field. This point of infinite curvature is known as a singularity. Singularities in physical theories often signal the limitations of the theory itself, suggesting that there might be more to the story. In simpler terms, the existence of singularities within the framework of General Relativity implies the need for a more comprehensive theory.

The appearance of singularities, particularly at the centre of black holes, has motivated physicists to explore alternative theories of gravity. Quantum theories of gravity, such as string theory, are among the candidates that may provide insights into the behaviour of spacetime at extreme scales. These theories propose a minimum distance scale below which the traditional understanding of spacetime curvature may not apply, potentially avoiding the appearance of singularities and paving the way for a more complete description of the gravitational realm.

The existence of black holes, while captivating, raises questions about the nature of these enigmatic entities and the validity of current theories describing them. Although the quest for a comprehensive understanding of black holes continues, the observational evidence supporting their existence is compelling.

At the heart of our galaxy, and potentially at the centres of all galaxies in the Universe, scientists propose the presence of supermassive black holes. The credibility of this hypothesis gains support from meticulous observations of a star named S2, which orbits around the intense radio wave source known as Sagittarius A* situated at the galactic centre. S2's orbit has been precisely measured, revealing a remarkably short orbital period of just over 15 years, making it the fastest-known orbiting object and reaching speeds up to 2 percent of the speed of light.

By understanding the precise orbital path of S2, astronomers can calculate the mass of the object it orbits, leading to the staggering determination that Sagittarius A* possesses a mass of 4.1 million times that of our Sun. Given S2's closest approach to the object at a mere 17 light hours, it is evident that Sagittarius A* must be smaller than this distance; otherwise, S2 would collide with it. The only plausible explanation for cramming such an immense mass into a space less than 17 light hours across is the presence of a black hole.

The confidence in the existence of a supermassive black hole at the centre of the Milky Way has been further substantiated through the study of an additional 27 stars, collectively known as the S-stars. These stars, with orbits taking them in close proximity to Sagittarius A*, contribute to the growing body of evidence supporting the hypothesis of a giant black hole residing at the core of our galaxy. Ongoing observations and refinements in the study of these stellar phenomena continue to deepen our understanding of the role of black holes in shaping the dynamics of galactic centres.

Black holes stand as captivating enigmas in our understanding of the cosmos. While their existence is established through indirect observations, the intricacies within their event horizons remain shrouded in mystery. Despite the impossibility of direct encounters with black holes, the physics governing the space within the event horizon is undoubtedly fundamental to our comprehension of the universe.

These cosmic entities hold immense significance, acting as catalysts for the development of a new theory of gravity and a paradigm shift in our understanding of space and time. The theoretical framework provided by Einstein's General Relativity, although immensely successful, reaches its limits when confronted with the extreme conditions near a black hole's event horizon. The search for a more comprehensive theory to describe these phenomena is a frontier of scientific exploration.

A coveted goal in observational astronomy is the discovery of a pulsar in orbit around a black hole. Such a cosmic arrangement is believed to exist somewhere in the vastness of the cosmos. Observing the behaviour of a pulsar – a highly precise cosmic clock – in the intensely curved spacetime proximate to a black hole holds the potential to push the boundaries of General Relativity to their limits. In the fortunate event of detecting anomalies or deviations from expected

behaviours, it could serve as a pointer towards the formulation of a novel theory, providing deeper insights into the gravitational dynamics surrounding black holes. The quest for such groundbreaking observations remains a thrilling pursuit for astronomers, offering the prospect of unravelling the mysteries hidden within the gravitational realms of black holes.

⊥ ANATOMY OF BLACK-HOLE

At the heart of a black hole lies a region known as the singularity. This is a point in space where gravity becomes infinitely strong, and the laws of physics as we know them break down. The singularity is shrouded by an invisible boundary called the event horizon. Once an object, including light, crosses this boundary, it is inexorably pulled towards the singularity and can never escape.

Surrounding the singularity is the ergo sphere, a region where the black hole's rotation drags spacetime along with it. This effect, known as frame-dragging, is a consequence of the extreme gravitational forces at play. Objects within the ergo sphere are forced to rotate in the same direction as the black hole, and escape becomes even more challenging.

The event horizon and ergo sphere together form the outer layers of a black hole, defining its boundary. As matter and energy are drawn into the singularity, they spiral inwards, creating a glowing accretion disk. This luminous disk is composed of superheated gas, dust, and other celestial debris swirling around the black hole at tremendous speeds.

Black holes come in different sizes, from stellar-mass black holes, formed by the collapse of massive stars, to supermassive black holes, found at the centres of galaxies and containing millions or even billions of times the mass of our Sun.

The anatomy of a black hole is a fascinating interplay of gravity, spacetime distortion, and the mysterious properties of matter under extreme conditions. While we can observe the effects of black holes on their surroundings, the interior, including the singularity itself, remains a realm of theoretical exploration, pushing the boundaries of our understanding of the universe.

In summary, a black hole is a celestial object defined by the presence of an event horizon, beyond which nothing, not even light, can escape the gravitational pull of the singularity at its core. The singularity, a point of infinite density, marks the breakdown of classical physics. Surrounding it, the ergo sphere exhibits the profound effects of the black hole's rotation on spacetime. As matter succumbs to gravitational forces, it forms a luminous accretion disk, showcasing the violent interplay between mass and spacetime distortion. Black holes vary in size, from stellar to supermassive, influencing their impact on the cosmos.

4. FORTUNE

The concluding chapter, **Fortune,** in the book **Marvels of The Cosmos** weaves a narrative that transcends time, bridging the distant past and the far-reaching future. Here, the reader encounters the fascinating intersection of engineering and cosmology, as the chapter explores the profound impact of nineteenth-century steam engine efficiency considerations on the birth of thermodynamics. Rather than an intentional quest to unveil the mysteries of the cosmos, the science of thermodynamics emerges as an unforeseen compass guiding our understanding of the universe's ultimate fate.

Within the pages of **Fortune,** the reader is taken on a journey through the intricacies of thermodynamics. This quintessentially nineteenth-century science, rooted in the efficiency of steam engines, unexpectedly provides a tangible foundation for speculating on events that will unfold countless eons from now—an astonishing timeline that stretches into an almost inconceivable 10^{60} years. The chapter's exploration of this vast expanse showcases the remarkable foresight of the pioneers of the steam age and underscores the enduring relevance and incredible reach of their scientific endeavours in shaping our understanding of the cosmos.

+ TIME'S PASSAGE

The narrative begins by highlighting the intrinsic nature of time, an indispensable concept so deeply embedded in the fabric of the universe that envisioning an existence without it is unfathomable. Despite its fundamental importance, modern science grapples with the challenge of fully elucidating this enigmatic property. Time, though intimately experienced by humans, is not a human creation. It serves as the regulating force shaping our days, and its inexorable passage propels our lives forward, marking our individual beginnings and inevitable ends.

Remarkably, the exploration of time transcends the human experience, extending to the entire cosmos. Time becomes intricately woven into the very essence of the universe itself. The narrative reflects on the dual nature of time, impacting both human evolution and the broader cosmic tapestry. In acknowledging our incomplete understanding, the narrative underscores the profound impact of our exploration of time. Through this endeavour, humanity has achieved something extraordinary – not only peering into the cosmic origins but also daring to contemplate the potential end of the universe.

The story invites contemplation on the symbiotic relationship between our comprehension of time and the mysteries of the natural world. It highlights the transformative power of investigating time's nature, unravelling the secrets of Earth's reality. The narrative suggests that the mere act of delving into the intricacies of time has granted humanity the ability to envision the universe's grand narrative, from its mysterious inception to the speculative realm of its eventual conclusion. The exploration of time emerges as a conduit to unravelling the profound mysteries that shape our understanding of existence and the cosmos.

Nestled on the arid coastal plain of northwestern Peru, the often-overlooked marvel of Chankillo reveals itself as one of South America's most captivating astronomical secrets. This ancient site, shrouded in mystery, tells the tale of a civilization flourishing two and a half millennia ago in an inhospitable landscape. The city, constructed by a civilization largely unknown to us, stands as a testament to their ingenuity and resilience.

Central to Chankillo's architectural prowess is its grandest structure—an imposing temple fortified atop a hill. Its walls, once adorned with brilliant white and adorned with red-painted figures, would have commanded attention against the backdrop of the vast desert. Today, the remnants of the temple's grandeur are diminished, with only faint traces of the original decorations surviving the passage of centuries.

The enigma of Chankillo deepens as archaeologists grapple with the puzzle of its strategic hilltop location. While the elevated site offers a commanding view across the desert, it raises questions about its defensive efficacy compared to other locations in the vicinity. Scholars posit that the true significance of Chankillo might not solely rest on the hill's summit but extend to the desert plain below.

Recent research has unveiled a new layer of understanding, proposing that the key to unlocking the mysteries of Chankillo lies in the desert expanse beneath the hill. This revelation challenges traditional assumptions, suggesting that the purpose and significance of the site may extend beyond its hilltop fortress. As archaeologists delve into the secrets buried beneath the arid plains, Chankillo emerges not just as a relic of the past but as a living testament to the ancient civilization's profound connection with the cosmos and the strategic intricacies that once governed their existence.

Beyond the weathered remnants of the fortress, an intriguing alignment of thirteen towers unfolds along the ridge of a nearby hill at Chankillo. Recent excavations have unearthed additional structures to the east and west of these towers, prompting archaeologists to reconsider the purpose of this reptilian-like arrangement. The intricate connection between these newly discovered buildings and the enigmatic reptilian structure unveils a story that transcends conventional assumptions.

To grasp the profound significance of this alignment, one must stand at the western observation point during the quiet hours before dawn, facing the brightening eastern horizon through the thirteen towers. As the night yields to the approaching day, a Chankillo dawn unveils a spectacle of unparalleled drama and evocation. The solar disc, its edge reddened and distorted by the dense desert air, abruptly materializes between two of the towers on the hill. In that fleeting moment, the sun transforms into a single, sparkling diamond suspended in the vast desert sky.

Within the span of seconds, the nearly imperceptible rotation of our planet begins to assert itself, dragging the celestial orb into full view. The brilliance of the rising sun commands attention, prompting an instinctive need to avert one's gaze, as if in reverence or to shield oneself from staring into the face of a god. The celestial ballet enacted through the alignment of towers at Chankillo unfolds as a testament to the ancient civilization's deep understanding of astronomical phenomena, intertwining the terrestrial and celestial realms in a mesmerizing display that continues to captivate observers across millennia.

The Thirteen Towers of Chankillo transcend mere architectural marvels; they embody an ancient calendar, silent witnesses to the passage of time for thousands of years, persisting long after their creators faded into the annals of history. This temporal chronicle eschews intricate clockwork mechanisms, opting instead for a reliance on the Sun, the most steadfast cosmic pulse available to the ancients.

In a breathtaking display of astronomical ingenuity, each of the thirteen towers serves as a meticulous timekeeper, strategically positioned to track the Sun's journey across the eastern horizon. On the Southern Hemisphere's summer solstice, December 21, the longest day of the year, the Sun ascends just to the right of the southernmost tower. This marks the commencement of its celestial odyssey, tracing a path across the horizon as Earth orbits the Sun. As the year unfolds, the sunrise gracefully progresses along the towers, reaching its zenith on June 21, the shortest day, as it appears just to the left of the northernmost tower.

This celestial dance allows the inhabitants of Chankillo to discern the date with remarkable precision at any time of the year, accurate within a range of two or three days. Standing at the western observing point on September 15, I marvelled at the Sun's consistent rise between the fifth and sixth towers—a celestial performance echoing across two millennia. Chankillo's enduring functionality as a calendar stem from the constancy of the Sun's rising and setting positions on the horizon, preserving the ancient wisdom encoded in these stones for generations and affirming the profound connection between humanity and the cosmos.

In the face of a breathtaking sunrise at Chankillo, set against the backdrop of a dramatic and tranquil landscape, one can't help but comprehend why the ancients likely deified the Sun. Despite a contemporary understanding of the Sun's scientific nature, the sheer magnificence of the rising orb in this sacred setting evokes a profound sense of reverence. The people of Chankillo, encountering such awe-inspiring moments, would have undoubtedly imbued the Sun with a divine essence.

Chankillo's significance extends beyond its utilitarian function as a calendar; it is a monumental complex that transcends mere timekeeping. Its grand scale, surpassing the pragmatic requirements of a calendar, reflects a multifaceted purpose. Chankillo emerges as part-clock, part-temple, and part-observatory—an architectural marvel where, on sacred occasions, the inhabitants could ceremoniously greet the appearance of their god, the rising Sun, amidst a spectacular setting.

The enduring grandeur of the Thirteen Towers of Chankillo attests to the elevated status accorded to this site by its ancient creators. These structures, with their intricate alignment and purposeful design, symbolize an ancestral instinct and desire to not only quantify but also understand the rhythm of the cosmic clock. Today, as the towers persistently cast shadows and trace the celestial movements, they stand as tangible testaments to the timeless pursuit of knowledge and the profound connection between humanity and the cosmic forces that have captivated our collective imagination for millennia.

✦ COSMIC CLOCK

Each day unfolds to the rhythmic cadence of our planet, a celestial dance that emanates from Earth's relentless spin at a staggering speed of over 1,500 kilometres (932 miles) per hour. This cosmic ballet, propelled by the Earth's axial rotation, orchestrates our existence by ushering us in and out of the Sun's luminous embrace. Earth's ceaseless motion, an unyielding force, imposes its temporal beat on the canvas of our lives with unwavering regularity.

The heartbeat of this planetary symphony is the day, defined by the twenty-four hours it takes for Earth to complete a full rotation on its axis. It manifests as the 86,400 seconds it demands for anyone stationed at the Equator to complete a journey around the Earth's substantial circumference, measuring 40,074 kilometres (24,901 miles). This rhythmic procession, an intricate manifestation of Earth's spin rate, serves as the most conspicuous and fundamental temporal cycle.

The origin of this perpetual rhythm is embedded in Earth's geological history, dating back 4.5 billion years to the planet's formation. As a rocky, iron-cored celestial sphere, Earth's axial rotation became a cornerstone of its identity, imprinting the unfolding narrative of time on its surface. The profound continuity of this cosmic rhythm underscores the enduring legacy of Earth's geological evolution and its role as the choreographer of our daily existence.

In our cosmic voyage, we hurtle through space at an astonishing speed of 108,000 kilometres (67,108 miles) per hour, engaged in an intricate dance around our life-giving star, the Sun. This celestial waltz unfolds at an average orbital distance of 150 million kilometres (93 million miles), defining the vast elliptical path that shapes our cosmic journey. Over the course of our cosmic sojourn, we complete one lap of this monumental track, covering a staggering 970 million kilometres (600 million miles).

The rhythmic cycle of our orbital journey meticulously marks the passage of time, epitomized by the concept of a year. This temporal journey spans 365 days, five hours, 48 minutes, and 46 seconds—a precise duration that encapsulates the Earth's orbital period around the Sun. In this celestial choreography, we return consistently to an arbitrarily designated starting point, perpetuating the cyclical nature of our annual pilgrimage through the cosmic expanse.

As we sweep through the cosmic arena in our orbit around the Sun, we both mark and measure the beginning and end of what we recognize as a year. This astronomical voyage not only delineates the fundamental structure of our timekeeping but also connects us intimately to the cosmic rhythms that govern our existence, underscoring the intricate interplay between our planet and the radiant star that graces our solar system.

Gazing into the celestial expanse, we encounter an array of celestial clocks, each marking the passage of time in distinctive rhythms. The moon, Earth's celestial companion, exhibits a captivating dance as it orbits our planet every 27 days, seven hours, and 43 minutes. Intriguingly, the moon is tidally locked to Earth, meaning it takes nearly the same amount of time to complete a rotation on its own axis—27 Earth days. This synchronization results in the moon consistently presenting the same face to Earth, a perpetual dance frozen in celestial harmony.

Venturing further into our solar neighbourhood, the red planet, Mars, mirrors Earth's daily cadence with a Martian day lasting one Earth day and an additional 37 minutes. However, the increased distance from the Sun elongates the Martian year, requiring 687 Earth days for Mars to complete a full orbit around our star. In the outer realms of the solar system, the temporal tapestry expands, with distant Neptune taking over 60,000 Earth days, equivalent to 165 Earth years, to complete its orbit around the Sun. Notably, in September 2011, Neptune celebrated its first full orbit since its discovery in 1846, providing a poignant testament to the vast scales and enduring regularities that characterize the cosmic timekeeping embedded in our celestial surroundings.

In our cosmic exploration, the clockwork of the cosmos unfolds with unwavering precision, and as our gaze extends into the vastness of space, the cycles become truly monumental, repeating on humbling timescales. While Earth and its planetary companions mark the passage of years with their orbits around the Sun, our entire solar system embarks on a colossal journey, tracing out its orbit around the galactic centre.

Amidst the cosmic tapestry, our solar system is but one among at least 200-billion-star systems in the Milky Way galaxy. Each of these stellar communities undertakes its own unique voyage around the galactic centre, and collectively, they all orbit the super-massive black hole residing at the heart of our galaxy. This immense journey takes approximately 225 million years, with our solar system hurtling through space at an astonishing speed of 792,000 kilometres (492,125 miles) per hour to complete one circuit—an epoch known as a galactic year.

Over the span of Earth's existence, which dates back four and a half billion years, our planet has completed a remarkable 20 revolutions around the galaxy. Hence, Earth is now 20 galactic years old. In the comparatively brief period since humans first appeared on Earth a quarter of a million years ago, less than one-thousandth of a galactic year has elapsed. In the grand scale of the cosmos, this fraction of time is equivalent to the fleeting length of a summer's afternoon in Earth's terms, underscoring the vastness and profound endurance inherent in the cosmic cycles that envelop our existence.

When we consider the entirety of our species' history to be the blink of an eye across the galaxy, this enormous span of time becomes impossible to understand. Since we experience history in minute, day, month, and year cycles, it is nearly hard to stretch our sense of time across a galactic year. Nonetheless, some species on Earth have lived for eons, spanning these most magnificent cycles.

✦ GALACTIC CLOCK

The ceaseless motion within the Universe ensures that nothing remains static; the galactic clock perpetually ticks, ushering everything into new chapters within the vast narrative of the cosmos. This cosmic timepiece diligently marks the passage of days, weeks, months, and years on each of the myriad planets scattered throughout our galaxy. In this cosmic symphony, time unfolds using its own inherent rhythms, guiding celestial bodies through their respective journeys and cycles.

As one ventures to the farthest reaches of our solar system, the length of a year expands, and the celestial cycles assume a grander scale. The diverse solar systems, numbering around 200 billion within our galaxy, each embark on their unique odyssey around the galactic centre. All these stellar communities, including our own, orbit the supermassive black hole that resides at the heart of the Milky Way Galaxy, a gravitational anchor shaping the collective destiny of countless stars and planets.

The intricate dance of the cosmos reveals that every solar system, with its planets and celestial entourage, contributes to the symphony of galactic motion. Each of these star systems, with its unique set of planetary orbits and cosmic phenomena, charts its own course around the gravitational epicentre. This journey around the galactic centre, orchestrated by the interplay of gravitational forces, encapsulates the grandeur and complexity inherent in the cosmic narrative.

In contemplating this cosmic ballet, one realizes that time, in its cosmic essence, transcends the human scale. The relentless progression of days, weeks, months, and years on every planet mirrors the broader cosmic rhythms, with each celestial body contributing to the intricate choreography of the galactic dance. The galactic clock, forever ticking, serves as a testament

to the dynamic nature of the Universe, guiding all celestial entities on their profound journeys through the cosmic tapestry.

✚ ANCIENT LIFE

The Ostional Wildlife Refuge, nestled on the Pacific coast of Costa Rica, unfolds as a sanctuary for one of nature's most breathtaking spectacles. Throughout numerous nights of the year, a select group of tropical beaches along this slender land bridge connecting North and South America bear witness to the arrival of prehistoric creatures. Emerging from the vast expanse of the ocean, these enigmatic beings  venture onto the shore to fulfil a primal instinct – the ancient ritual of laying their eggs in the warm embrace of the sand.

Our filming expedition took us to Playa Ostional, a diminutive stretch of sand that shares its proximity with a welcoming village huddled around an improvised football pitch. This unique locale stands out as one of the rare beaches worldwide where a significant number of sea turtles choose to create their nests. The events that unfold in this coastal haven represent a pivotal chapter in one of the oldest life cycles on Earth.

As the waves caress the shores of Playa Ostional, sea turtles, guided by an instinct passed down through countless generations, embark on a journey from the ocean depths to the sandy shores. Here, amidst the tranquillity of the refuge, they laboriously dig nests to safeguard their precious cargo—their eggs. This extraordinary spectacle is a testament to the ancient rhythms of life that persist in this ecological haven, reminding us of the enduring cycles that connect us to the distant epochs of Earth's past. The juxtaposition of the makeshift football pitch and the timeless nesting grounds of these prehistoric creatures highlights the harmonious coexistence of human settlements and the remarkable biodiversity that thrives in the Ostional Wildlife Refuge, making it a haven where nature's grandeur continues to captivate and inspire.

In the heart of our filming expedition at the Ostional Wildlife Refuge, our focus lies on capturing a timeless spectacle—the ritualistic journey of turtles from the expansive ocean to the sandy shores, a tradition meticulously preserved for over 120 million years, spanning half a galactic year. Armed with night-vision camera equipment, we patiently await the emergence of these ancient creatures, pondering the profound historical disparity between their existence and the comparatively brief timeline of the species responsible for building the adjacent football pitch.

As the night unfolds, it's impossible not to contemplate the colossal mismatch in histories. The turtles, with a lineage dating back to a time when continents bore unfamiliar configurations, have persevered through the eons, awaiting the opportune moment to embark on their pilgrimage to lay eggs in the sand. Their ancestors navigated oceans framed by coastlines and landmasses that would appear unrecognizable to our contemporary understanding of Earth's geography. North America once nestled close to Europe, South America embraced Africa, and Australia stood united with Antarctica.

In awe, we witness these ancient creatures, descendants of an epoch when the very shape of continents was in flux, carefully excavating the sand to create nests for their precious eggs. The turtles' ritual, unchanged across epochs, unfolds against the backdrop of a planet that has undergone transformative shifts over geological timescales. It's a poignant reminder of the turtles' enduring connection to the evolving Earth, a connection forged long before humans marked out familiar territories.

The turtles' patient vigil, undertaken as continents drifted and landscapes transformed, offers a glimpse into the profound temporal mismatch between their ancient lineage and our relatively recent presence on Earth. As they tenderly cover their eggs, an act of care repeated over countless generations, it becomes apparent that these ancient beings have borne witness to the reshaping of our planet and the celestial dance above. The patterns of stars, visible from their vantage point on the other side of the Galaxy, reflect a perspective vastly different from our own. In this moment of silent return to the ocean, each turtle becomes a living testament to the enduring cycles of nature and the vastness of time that transcends human comprehension.

⊥ TIME MEASUREMENT

The human journey in measuring time has spanned millennia, evolving from primitive methods to the precise temporal measurements achieved in contemporary times. The inception of chronometry traces back approximately thirty thousand years ago, to the Stone Age, where early humans harnessed the lunar cycle as a rudimentary tool for marking time. The Moon, with its consistent and observable phases, emerged as a prominent celestial guide, offering a clear rhythm in the night sky to our ancient ancestors.

For these early humans, attuned to the celestial theatre, the lunar cycle provided a fundamental framework for their first attempts at chronology. By keenly observing the waxing and waning of the Moon, they fashioned rudimentary calendars. This lunar-based chronometry extended beyond merely tracking the day-night cycle, providing a broader structure to the passage of time throughout the year. The naming of distinct periods, a result of this lunar observation, marked the initial steps in humanity's classification and division of the cosmic cycles.

Through their connection with the lunar phases, early humans bestowed order upon the natural progression of time. The lunar calendar, shaped by the apparent movements of Earth's celestial companion, enabled them to distinguish between various intervals, creating a nascent system that laid the foundation for more intricate temporal measurements. As our ancestors gazed at the night sky, deciphering the celestial rhythms, they unknowingly initiated a journey of discovery that would eventually lead to the sophisticated timekeeping mechanisms we employ today.

The precise division of the day necessitated the development of one of our most enduring technological innovations, the impact of which has been immeasurable, in addition to the designation of the morning, afternoon, and evening.

The earliest forays into timekeeping involved simple yet ingenious devices, with sundials taking centre stage across various ancient civilizations. Utilizing a basic stick known as a 'gnomon' to cast a shadow, sundials enabled people to track the passage of time during daylight hours by measuring the movement of the shadow across a calibrated surface. Despite their surprising accuracy, sundials had limitations, particularly their reliance on sunlight, rendering them impractical on cloudy days and entirely unusable at night.

Ancient Egypt stands out as a pioneering civilization that transcended the constraints of sundials. Around 1400 BC, during the reign of Pharaoh Amenhotep III, Egypt introduced a groundbreaking technique: the water clock. Although the concept of using water to measure time might date back to around 6000 BC, the oldest physical evidence of a water clock is traced to this Egyptian era. These early water clocks were elegantly designed stone vessels that facilitated a near-constant escape of water through a hole in the base. Inside the vessel, twelve markings allowed for time measurement as the water level gradually dropped. These innovative devices provided accurate timekeeping day and night, enabling priests to conduct their rituals at precise hours regardless of sunlight.

The use of water clocks persisted and evolved across diverse cultures for centuries. Concurrently, hourglasses emerged as another timekeeping tool, leveraging the flow of sand to measure time. An intriguing historical example is the Portuguese explorer Ferdinand Magellan, who employed 18 hourglasses as navigation tools on his ship during the historic circumnavigation of the globe in 1522. These hourglasses served as vital instruments for ensuring accurate time measurements and aiding navigation during the ambitious maritime expedition.

The development and utilization of these early timekeeping devices mark significant milestones in humanity's quest for precision in measuring time, showcasing the resourcefulness and ingenuity of civilizations across different eras and regions.

The advent of pendulum clocks marked a revolutionary leap in the accuracy of timekeeping, and it was the pioneering work of Galileo that laid the foundation for understanding the physics behind the swinging pendulum. The key property that renders a pendulum effective for timekeeping is its consistent period of swing—the familiar tick-tock sound of a clock—and this period is solely contingent on the length of the pendulum and Earth's gravitational pull.

In a seemingly counterintuitive manner, the period of the pendulum's swing remains independent of how high it is initially lifted, as long as the lift is not excessive. This characteristic adds to the reliability of pendulum clocks. Physics students often commit to memory the formula that governs the time period of a pendulum, etching it permanently in their minds:

$$T = 2\pi\sqrt{\frac{L}{g}}$$

Here, **T** represents the period of the pendulum's swing, π is a mathematical constant (approximately 3.14159), **L** signifies the length of the pendulum, and **g** denotes Earth's gravitational acceleration.

This formula encapsulates the fundamental relationship between the length of the pendulum, the gravitational force acting upon it, and the resulting period of oscillation. The simplicity and precision of the pendulum clock, governed by the principles elucidated by Galileo, transformed timekeeping, offering an unparalleled level of accuracy and becoming a cornerstone in the development of precise clock mechanisms.

In modern times, the pinnacle of timekeeping precision is achieved through the application of atomic clocks, which rely on the fundamental properties of atoms, typically those of caesium. The remarkable accuracy of atomic clocks stems from the use of the frequency of light emitted when electrons undergo transitions within atoms, serving as the analogue to the traditional pendulum in classical clocks.

The unchanging nature of atomic structures ensures a constant and reliable frequency of emitted light. Ingenious engineering utilizes this characteristic to keep an oscillator ticking at an exceptionally precise rate, allowing atomic clocks to measure time with an astonishing accuracy of one-thousand-millionth of a second per day. The defining feature of atomic clocks is their ability to maintain consistency by leveraging the stable properties of atoms, providing an unparalleled standard for precision timekeeping.

The very definition of the second, established since 1967, embodies the principles underpinning atomic clocks. A second is officially defined as the duration of 9,192,631,770 periods of the radiation corresponding to the transition between the two hyperfine levels of the ground state of the caesium 133 atom. In simpler terms, this means that a second is the time it takes for 9,192,631,770 peaks in a wave of light, emitted during a specific electron transition within an atom of caesium, to pass by a given point.

This definition crystallizes the intrinsic connection between the physical world and timekeeping precision. By anchoring the definition of a second to the behaviour of caesium atoms, atomic clocks offer an unerring standard for measuring time, underlining the impressive intersection of fundamental physics and practical timekeeping technology. The legacy of atomic clocks underscores how our understanding of the atomic realm has been harnessed to refine and redefine our grasp of time, providing a bedrock for myriad scientific and technological applications.

✛ TIME'S ARROW

The Perito Moreno glacier, nestled in the breathtaking landscape of Patagonia in southern Argentina, stands as a natural wonder and a testament to the awe-inspiring beauty of our planet. Situated within the expansive Los Glaciares National Park, this dense blue behemoth of frozen water is a pivotal component of a glacier system originating from the southern Patagonian ice fields, collectively forming the third-largest icecap on Earth.

The sheer magnitude of the Perito Moreno glacier is staggering. Covering an expansive area of 250 square kilometres (96 square miles), and in some regions reaching a depth of 170 meters (560 feet), it is an imposing testament to the forces of nature.

The glacier, like a colossal frozen river, terminates where solid ice meets liquid water at Lake Argentino, creating a mesmerizing spectacle. Here, a great wall of ice rises majestically over the lake's surface, forming a dramatic panorama that captivates those fortunate enough to witness it.

Sailing along the edge of the Perito Moreno glacier offers an unparalleled experience. The fortunate few who venture to this remote yet utterly captivating location are treated to a journey across one of the most dramatic expanses of water in the world. The glacier's dense blue hues, contrasting against the backdrop of the surrounding mountains and pristine wilderness, create a visual symphony that is both stark and sublime.

This pristine and seemingly inhospitable terrain serves as a sanctuary for unique flora and fauna adapted to the harsh conditions, adding to the ecological significance of the region. The Perito Moreno glacier is not merely a geological marvel; it's a living entity that shapes the landscape, influences local ecosystems, and provides a glimpse into the profound forces at play in the natural world. Visiting this remote and remarkable place is a journey into the heart of Earth's majestic landscapes, where the powerful dance between ice and water unfolds in a spectacle of unmatched beauty.

Upon first glance, the Perito Moreno glacier exudes an aura of stillness and timelessness. Standing on the shores of the lake, it seems like a realm where time's passage goes unnoticed, adhering to the laws of physics in a manner that allows for an almost surreal serenity. However, as one approaches, a profound realization sets in – this glacier is far from static. It is a dynamic force in perpetual motion, an ancient entity relentlessly carving its way down from the Andes, a process ongoing for tens of thousands of years.

At the glacier's edge, an imposing ice cliff, reaching a height of 70 meters (230 feet), commands attention. The entire face of the glacier is engaged in a constant journey, sliding into the lake at an impressive rate of around 50 centimetres (20 inches) per day. This means that over a quarter of a billion tonnes of ice cascade into the lake annually, a spectacle that transcends visual appreciation to invoke a profound sense of the glacier's active existence.

Approaching the glacier is not only a visual experience but a visceral one. The glacier is alive with movement and sound. Periodically, a tremendous cracking sound reverberates through the air, followed by a deep rumbling that hints at the immense forces at play. The surface of the lake springs to life as a turbulent wave surge beneath any boat in its vicinity. The glacier's metamorphosis is not a gradual process; it's a spectacle of transformative power, challenging the perception that changes in such vast landscapes occur at a glacial pace.

The Perito Moreno glacier, with its intricate dynamics and colossal movements, defies any notion of stillness. It is a living, unpredictable, and overwhelmingly powerful entity, an ancient force of nature that continuously reshapes the landscape as it inexorably slides into the waters. Standing in its presence, one feels the pulse of time and the unstoppable march of nature, making the Perito Moreno glacier a testament to the profound dynamism inherent in Earth's geological wonders.

The unfolding drama of the Perito Moreno glacier is intricately woven into a highly ordered sequence, a choreography of natural events that reflects the perpetual passage of time. This sequence, governed by the laws of nature, involves the rhythmic cycle of snowfall, ice formation, and the glacier's gradual descent down the valley. As the ice meets the water, the culmination of this sequence occurs – pieces break off, cascading into the lake and generating waves in a visually striking display.

This ordered progression of events encapsulates a fundamental aspect of our perception of time – the expectation that events transpire in a specific sequence. Witnessing ice fall from the glacier, splash into the water, and create waves is not merely an observation but a visceral confirmation of the regularity inherent in natural processes. The predictability of these events aligns with our intuitive understanding of the world – a world where sequences unfold in a certain order, and deviations from this order would signal something amiss.

However, the concept of events happening 'in order' prompts a legitimate question about the nature of this order. While standing by the edge of the lake, we intuitively comprehend the unlikelihood of witnessing this dramatic sequence in reverse. There is no inherent physical barrier preventing water molecules from gathering on the surface, reducing temperature, and forming ice that leaps onto the glacier. Yet, despite the absence of a physical constraint, we recognize the implausibility of such a reversal.

The explanation for this asymmetry in our experiences of time lies in the concept known as 'time's arrow.' While the laws of nature do not prohibit events from occurring in reverse, the overwhelming statistical improbability of such occurrences, coupled with the increase in entropy, provides a scientific basis for our observation that certain sequences of events tend to unfold in one direction – the direction of increasing disorder and entropy, in alignment with the arrow of time. The Perito Moreno glacier, with its cyclic dance of ice and water, serves as a vivid canvas on which the narrative of time unfolds, each sequence a testament to the ordered yet dynamic nature of our universe.

The concept of "time's arrow" was introduced by the British physicist Sir Arthur Eddington in the early twentieth century, representing a seemingly simple yet profound quality that governs the directionality of time in our universe. Eddington, a key figure in bringing Einstein's theory of relativity to the English-speaking world during World War I, played a pivotal role in directly confirming the theory's findings. His leadership in the 1919 solar eclipse expedition solidified his reputation as a prominent scientist.

In his 1928 publication, "The Nature of the Physical World," Eddington introduced two enduring ideas that have left an indelible mark on popular scientific culture. The first was the vivid metaphor of the infinite monkey theorem, suggesting that given an infinite amount of time, anything consistent with the laws of physics is bound to happen. He playfully expressed this idea with the notion that an army of monkeys, given typewriters and infinite time, could eventually produce all the books in the British Museum.

Eddington's second enduring idea was the concept of "time's arrow," a metaphorical representation of the one-way property of time that lacks an analogue in space. In explaining this concept, he proposed a simple yet insightful thought experiment. Imagine arbitrarily drawing an arrow to represent the progression of time. As one follows the arrow, if there is an increasing presence of randomness or disorder in the state of the world, it signifies that the arrow points toward the future. Conversely, if the random element decreases, the arrow points toward the past. Eddington's contention was that the introduction of randomness is the irreversible factor in the arrow of time, a quality that cannot be undone.

This concept resonates with the fundamental understanding in physics that the increase in entropy, or disorder, is a fundamental characteristic of the arrow of time. Eddington's articulation of "time's arrow" has endured, providing a conceptual framework to comprehend the asymmetry of time's flow and the unique irreversibility associated with the unfolding of events in the universe.

The fact that Eddington's arrow only points in one direction effectively and graphically conveys a fundamental aspect of time. By randomness, though, what does he mean? The universe is clearly always changing, but why is this evolution occurring? How do we measure the degree of randomness in something? What distinguishes the past from the future? Why does time have an arrow?

✚ ORDER OF DISORDER

In 1712, Sir Thomas Newcomen, an English inventor, created the first commercially successful steam engine, a significant technological advancement that would set the stage for the Industrial Revolution. However, the true transformation of this revolutionary technology was propelled further by the innovations of the Scottish inventor James Watt. In 1763, Watt was tasked with repairing a Newcomen engine by the University of Glasgow, an endeavour that would lead him to develop a groundbreaking steam engine, altering the trajectory of modern life.

Watt's steam engine surpassed its predecessor in efficiency and flexibility. It consumed significantly less coal for a given power output, translating into lower operational costs. Moreover, Watt's engine exhibited the capacity to generate rotary motion, a crucial capability for powering machinery on the factory floor. Unlike earlier engines that were constrained to locations near rivers for their water-pumping function, Watt's engine allowed factories to be situated anywhere, catalysing the emergence of the modern industrial landscape.

The advent of steam-powered machines, driven by Watt's innovations, played a pivotal role in reshaping history. Despite their transformative impact, nineteenth-century engineers faced challenges in improving steam engines beyond Watt's design. Fundamental principles seemed to limit their efficiency. Given the imperative to maximize profit margins, even a slight enhancement in effectiveness held significant value for businesses. Consequently, questions regarding engineering design, such as determining optimal fire temperature and the choice of substances for boiling in the engine, not only intrigued scientists but also became critical considerations for industrial enterprises.

It was within the context of these engineering challenges that the science of thermodynamics emerged. This scientific discipline, born out of the practical concerns of improving steam engines, introduced precise concepts of heat, temperature, and energy into the scientific vocabulary for the first time. The pursuit of more efficient steam engines led to foundational advancements in understanding the principles governing energy transformations, laying the groundwork for the development of thermodynamics and its enduring impact on science and technology.

One of the prominent figures engaged in addressing the challenges of steam engine efficiency was the German mathematician, Rudolf Clausius. In the first half of the nineteenth century, the prevailing understanding of heat considered it as a fluid that flowed from hot objects to cold ones. This conceptualization, however, proved inadequate in explaining the intricate workings of a steam engine cycle.

Rudolf Clausius, along with other scientists, recognized the limitations of the prevailing heat fluid model. A pivotal contributor to Clausius's theoretical advancements was James Joule, an English physicist and brewer. Joule, driven by the pragmatic goal of improving the efficiency of the steam engines in his brewery, embarked on a scientific journey that would significantly impact the understanding of thermodynamics. The motivation behind Joule's investigations was none other than the quest for cheaper beer production.

Joule delved into exploring the relationship between the work achievable by his steam engines and the concept of heat. In the pursuit of optimizing the brewing process, he managed to reduce the costs associated with beer production. In doing so, Joule inadvertently laid one of the foundational cornerstones of the science of thermodynamics. The unexpected intersection of brewing and fundamental physics exemplifies how practical challenges and everyday motivations can drive scientific inquiry and lead to groundbreaking discoveries.

James Joule conducted a series of elegantly simple experiments that provided empirical evidence for the conversion of mechanical work into heat, a groundbreaking contribution to the understanding of thermodynamics. In one such experiment, he ingeniously utilized a falling weight to set a paddle into motion within an insulated barrel filled with water. The falling weight's mechanical work was thus transformed into heat within the water.

Joule's meticulous approach involved measuring the amount of work performed by the falling weight and correlating it with the resulting temperature rise in the water. This experiment, along with others conducted on compressed gases and flowing water, consistently demonstrated that a fixed amount of work produced a uniform temperature increase of one degree Fahrenheit in a given quantity of water.

His tombstone, located in Brooklands cemetery near Manchester, bears the numerical inscription

772.55. This number represents Joule's measurement of the amount of work, expressed in foot-pounds force, required to elevate the temperature of one pound of water by one degree Fahrenheit. Joule's experiments and precise measurements laid a solid foundation for the principle of conservation of energy, linking mechanical work and heat in a manner that significantly contributed to the development of thermodynamics.

James Joule's pivotal work played a crucial role in reshaping our understanding of heat as he demonstrated that it is not a substance that can be created or destroyed but rather a form of energy. In his groundbreaking experiments, Joule illustrated that heat is a measure of energy, akin to potential energy stored in a resting ball on a table. This energy can be released by transferring it from a hot object to a cold one.

Joule's experiments, particularly those involving a falling weight to perform work and generate heat, highlighted that the method of doing work (whether by a falling weight, shining light, or an electric current) is inconsequential as long as the same amount of work is performed. This profound insight led to the quantification of these principles into the First Law of Thermodynamics. This law states that energy cannot be created or destroyed; it can only change from one form to another.

Rudolf Clausius, building on Joule's foundation, explicitly formulated the First Law of Thermodynamics in his seminal 1850 publication titled 'On the mechanical theory of heat.' This law is fundamental to our understanding of energy conservation and serves as a cornerstone in the science of thermodynamics. It signifies that the total energy within a closed system remains constant, and any energy change involves conversion between different forms.

The First Law of Thermodynamics, expressed mathematically as $\Delta U = Q - W$, states that the change in internal energy (ΔU) of a system is equal to the heat added to it (Q) minus the work done by it (W). The sign conventions indicate that if work is performed on the system, it contributes positively, and if heat is extracted from the system, it contributes negatively.

Clausius introduced the concept of entropy, a crucial element in the Second Law of Thermodynamics. This law, as stated by Clausius, asserts that "No process is possible whose sole result is the transfer of heat from a body of lower temperature to a body of higher temperature." This seemingly simple proposition carries profound implications for the future of the universe.

Entropy, denoted as **ΔS**, becomes central when the Second Law is expressed quantitatively. The change in entropy of a system, like a tank of water, is related to the amount of heat added to it at a fixed temperature (**ΔQ**). Symbolically, **ΔS = ΔQ/T**, where **T** is the fixed temperature. The crucial insight from Clausius is that in any physical process, entropy either remains the same or increases; it never decreases. This is the thermodynamic arrow of time.

Clausius discovered a measurable and quantifiable physical quantity (entropy) that consistently increases in practice and never decreases in theory. This insight has practical implications for designing systems like steam engines, setting fundamental limits on their efficiency. It also counters the idea of creating perpetual motion machines, emphasizing that you can't get something for nothing.

Eddington's cryptic quote about randomness and the arrow of time becomes clearer when we understand that entropy is equated with the amount of randomness in the world. The Second Law implies that entropy, or randomness, always increases, introducing a distinction between the past and the future. In the future, entropy will be higher, and in the past, it was lower. This understanding of entropy helps explain why the Second Law of Thermodynamics predicts the eventual heat death of the universe.

✚ ENTROPY AT WORK

In 1908, Zacharias Lewala, a railway worker in the small town of Kolmanskop in southern Namibia, discovered a single diamond in the sand. Recognizing the significance of this find, railway inspector August Stauch, his manager, initiated a series of events that transformed the desolate Kolmanskop into one of the world's most valuable diamond mines.

The German colonial government closed the area to outsiders, granting exclusive rights to German entrepreneurs to exploit the diamond wealth. Over the next 40 years, Kolmanskop flourished as a bustling community, attracting over a thousand people eager to strike it rich by extracting diamonds from the desert. The town became a symbol of German opulence in the heart of the arid landscape.

As wealth poured in, residents constructed a town reminiscent of grand German architecture. The landscape featured lavish houses, a casino, a ballroom, and even the first X-ray station in the Southern Hemisphere. Despite the challenging desert environment, the inhabitants led a champagne lifestyle, creating a unique blend of luxury amid the harsh desert surroundings.

However, like many mining booms, the prosperity of Kolmanskop was temporary. Over time, the diamonds became increasingly difficult to find, and the town gradually lost its lustre. By 1954, Kolmanskop was abandoned, and over the subsequent half-century, the once-thriving structures fell into disrepair, succumbing to the relentless advance of the desert sands. Today, the ghost town stands as a haunting reminder of its glittering past, with buildings slowly being reclaimed by the shifting sands of time.

Kolmanskop, now a ghost town, stands as a testament to the challenges posed by the harsh environment and relentless winds of the southern Namibian coast. Located just outside the modern port town of Lüderitz, which sits in splendid isolation along the southern Namibian coast, Kolmanskop's attempt to replace the natural grandeur of the desert with human-made architectural wonders has been thwarted by the power of the winds.

Lüderitz is known for its particularly harsh conditions, even by the standards of this arid region, mainly due to the relentless winds from the South Atlantic. These winds pick up fine-grained sands from the Namib Desert and relentlessly propel them into everything in their path, from machinery and houses to camera equipment and eyes. The wind in this region is so forceful that it can make walking nearly impossible without shielding one's face with hands. The abrasive sand-laden wind is a constant challenge for anyone in the vicinity.

Filming in Kolmanskop presented its own set of challenges. The winds were so fierce that the sand acted like a lacerating force, making it difficult for individuals to look directly into the wind without protection. Even camera equipment, left in the desert for extended periods to capture scenes, suffered the consequences. High-precision optics on the camera's lens were sandblasted, giving them a sandpaper-like texture after just a single afternoon in the vicinity of Lüderitz.

Despite the challenging conditions, Kolmanskop remains standing, defying the winds and preserving the remnants of a bygone era. If not for the arid climate where it rarely rains, preventing erosion and decay, the ghost town would likely have already succumbed to the relentless forces of the desert.

The relationship between entropy, randomness, and decay is eloquently illustrated by the small sandcastle that is gradually disintegrating in the desert wind. We'll need a definition of entropy that is distinct from Clausius's and far more intuitive in order to comprehend why this is the case. Ludwig Boltzmann created what is referred to be the statistical definition of entropy in the 1870s.

Boltzmann's definition of entropy provides a mathematical insight into the distinction between the arrangement of sand grains in a sandcastle and a sand pile. Entropy, denoted as **S**, is defined by the equation:

$$S = k_B \cdot \ln(W)$$

Here, k_B is Boltzmann's constant, and W represents the number of ways in which the constituent parts of a system can be rearranged without noticeable change. The natural logarithm ($\ln$) is used to relate the number of arrangements to entropy.

In the case of a sandcastle, the entropy is low because there are relatively few ways to rearrange the sand grains while maintaining the specific, highly-ordered structure of the castle. The sandcastle's distinct shape limits the number of arrangements that preserve its form.

Conversely, a sand pile has higher entropy because nearly any rearrangement of the sand grains will result in a pile, and one arrangement is indistinguishable from another when considering a general pile of sand. Therefore, the sand pile has a greater number of possible arrangements, leading to higher entropy.

Boltzmann's equation, engraved on his gravestone, encapsulates this concept, emphasizing the relationship between the number of possible arrangements and the entropy of a system. It offers a quantitative measure of disorder or randomness within a given system, illustrating the fundamental principle that systems tend to evolve towards states with higher entropy over time.

This may sound a little confusing and still not fully clear, but the important thing to remember is that if you start shifting the sand grains randomly, they are far more likely to form an empty pile of sand than a sandcastle, provided that each unique arrangement of the grains is equally likely. This is due to the fact that the majority of the arrangements you make at random resemble formless piles, and relatively few resemble sandcastles.

At the microscopic level, the increase in entropy becomes a matter of probability and statistical likelihood. Considering individual sand grains on the turrets of a sandcastle, there is no inherent physical law preventing the wind from blowing a grain off one turret and randomly placing it back or replacing it with another grain from the desert. However, the crucial factor is pure chance.

The fundamental principle here, expressed by Boltzmann's concept of entropy, is that the number of ways to rearrange the sand grains into a formless pile is significantly greater than arranging them into the structured form of a castle. Therefore, when the wind acts randomly on the sandcastle, it is statistically more likely to contribute to the disintegration of the castle rather than reconstructing its original form. This statistical preference for higher entropy configurations is a reflection of the sheer multitude of ways in which disorder can manifest compared to the limited ways in which order can be maintained.

The deep reason behind the inevitable increase in entropy over time is the intrinsic likelihood of more disordered states. Although there is no explicit law prohibiting a decrease in entropy, the overwhelming probability of entropy increase makes it highly improbable for a system to spontaneously evolve towards a more ordered state. It's akin to the probability of flipping a coin billions of times and getting heads every single time – an exceedingly rare occurrence.

Boltzmann's statistical definition of entropy, as reiterated in a slightly different way, provides profound insights into understanding the arrow of time, as articulated by Eddington. Imagine a scenario where there are a million different ways to arrange a handful of sand grains. Out of these, 999,999 ways lead to the formation of disordered sand piles, while only one arrangement results in the creation of a beautifully ordered castle.

Now, if you repeatedly throw the sand grains up in the air and let them fall randomly, the statistical likelihood is that they will land in the form of a disordered pile more often than in the form of an ordered castle. This inherent statistical preference for disorder is the key to understanding why, over time, a force like the wind, which acts to rearrange things, tends to make them messier or disordered. The crucial point is that there are vastly more ways for a system to be disordered than ordered.

In this context, there emerges a distinction between the past and the future. The past, having had less disorder and more order, is characterized by greater organization. Conversely, the future is expected to be less ordered and more disordered because the statistical likelihood favours an increase in disorder over time. This difference in the likelihood of ordered and disordered states is what Eddington referred to as the future being more random than the past, and it aligns with his concept of the arrow of time pointing in the direction of increasing randomness. Hence, entropy always increases as a natural consequence of the statistical tendencies governing the arrangements of particles or elements within a system.

The understanding of entropy and the arrow of time provides a framework for explaining the progression from order to disorder over time, as observed in the universe. However, a significant question arises when considering the origins of the ordered structures we see today, such as planets, stars, and galaxies.

While the concept of entropy highlights the tendency for systems to evolve from ordered to disordered states, it does not inherently explain why the universe began in a highly ordered state. For instance, how did Earth, with its intricate order, come into existence? What about the Milky Way, composed of billions of ordered celestial bodies orbiting billions of ordered stars?

The challenge lies in comprehending why the universe initiated with sufficient order to permit the emergence of complex structures. While gravity can create local order, as seen in the formation of solar systems and stars, it does so at the expense of creating more disorder elsewhere. Hence, the initial conditions of the universe must have possessed a substantial amount of order.

The crucial question remains: Why did the universe start in such a highly ordered state, considering that highly ordered states are less likely to occur by chance? The analogy is drawn to the creation of a sandcastle, where the winds of chance are less likely to form an ordered structure than a disordered pile of sand.

Considering that the universe is currently far less ordered than it was 13.75 billion years ago, one could speculate that it is more probable for the universe to have spontaneously appeared a billionth of a second ago in a fully formed state, complete with planets, stars,

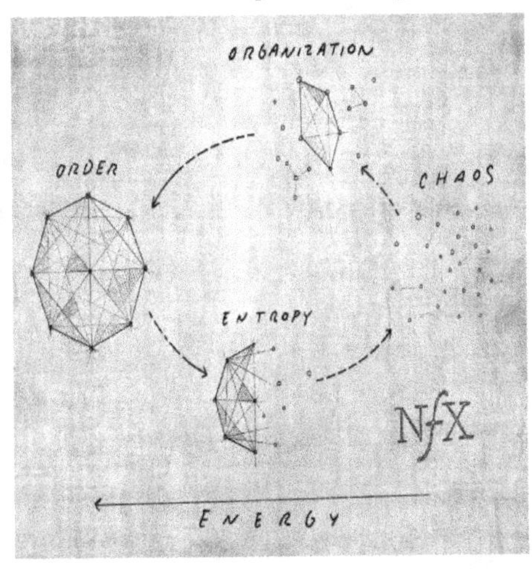

galaxies, and life, rather than originating from the Big Bang in a highly ordered state.

This raises a fascinating and unanswered question about the entropy of the early universe and the underlying reasons for its initial conditions, which scientists are actively exploring and researching to deepen our understanding of the cosmos.

⊥ LIFE OF THE UNIVERSE

The life of the universe unfolds in distinct stages, mirroring the cycles of birth, life, and death observed in individual entities like human beings, planets, and stars. The cosmic journey commenced approximately 13.75 billion years ago with the explosive event known as the Big Bang. In the initial phase, labelled the Primordial Era, the universe existed in a state devoid of the light emitted by stars. However, during these early years, the intense heat generated by the swirling matter would have caused the cosmos to radiate brightly, akin to a sun.

The violent conditions persisted for the first 100 million years, preventing the formation of stars. As the universe expanded and cooled over time, the feeble force of gravity gradually gained the upper hand. This allowed it to initiate the clustering of primordial elements such as dust, gas, and dark matter into the early structures of galaxies. This transformative period marked the onset of the second significant epoch in the cosmic timeline—the Stelliferous Era, aptly named the age of stars.

During the Stelliferous Era, galaxies began to host the formation of stars. These celestial bodies, fuelled by the gravitational collapse of cosmic materials, ignited in brilliant displays of light and heat. The emergence of stars signalled a profound shift in the character of the universe, introducing the dynamic interplay of stellar processes and the birth of planetary systems within galaxies.

The Stelliferous Era stands as a testament to the transformative power of gravity, shaping the cosmic landscape into a tapestry of luminous stars and galaxies—a captivating chapter in the ongoing narrative of the universe.

One of the most poignant turning points in the universe's history is the time the first stars were formed. It marks the conclusion of an extraterrestrial era in which there was no structure in the universe and it was just a formless nothingness. The era of light began at the beginning of the Stelliferous Era, which is also the point at which the universe would have started to resemble our own. The sky would have turned pitch-black, broken only by the piercing silver of the stars and the luminous mist of the galaxies. This is the universe we live in today, where the stars brighten our days and adorn our nights.

There are one hundred billion galaxies in the observable universe, of which our sun is one. Our galaxy has at least two hundred billion stars. Our universe is made up of an endless number of islands, each with its own starry canopy that fills it with light. We are only at the beginning of the universe, even though it has existed for more than 13 billion years. We are living toward the start of the Stelliferous Era, an era of astounding beauty and complexity, despite the fact that the universe is full of stars, massive nebulae, planet systems, and untold billions of worlds that we have yet to discover. However, the universe is dynamic and ever-evolving; this state won't last forever as time arrows out, it produces a cosmos that is as dynamic as it is beautiful.

On April 23, 2009, at 07:55 GMT, NASA's Swift satellite detected a gamma-ray burst, marking one of the most distant cosmic explosions ever observed. Gamma-ray bursts are brief, intense bursts of gamma-ray radiation, and they rank as the most energetic and powerful events in the known universe. These phenomena typically last only a few seconds and are believed to be associated with supernova explosions, representing the final stages of massive stars as they collapse to form black holes.

The Swift satellite, specifically designed and constructed for the study of gamma-ray bursts, swiftly responded to the detection. Within minutes of the burst, by 08:16 GMT, the UK's Infrared Telescope (UKIRT) in Hawaii captured the faint glow of the explosion's afterglow. The global astronomical community quickly joined the observation effort as the day progressed, with major telescopes around the world directing their focus toward the cosmic event as it became visible in their respective regions.

The afterglow of the gamma-ray burst was monitored for several hours, providing valuable data and insights into the characteristics of these distant and energetic phenomena. However, by April 28, 2009, the afterglow had faded entirely from view, concluding this brief yet intense episode in the cosmic landscape. The coordinated efforts of space and ground-based observatories during such events contribute significantly to our understanding of the universe's most extreme and dynamic events.

The unremarkable-looking entity is the fading remnant of GRB 090423, once among the brightest stars in the Universe. GRB 090423 was originally a Wolf-Rayet star, a class of massive stars with over twenty times the mass of the Sun. Due to their immense size and brightness, Wolf-Rayet stars have relatively short lifespans, burning out after a few hundred thousand years. When these massive stars exhaust their nuclear fuel, they undergo a rapid collapse, releasing an extraordinary amount of energy in a brief moment, surpassing the total energy output of our Sun over its 10-billion-year lifetime.

GRB 090423, with an estimated mass of 40 to 50 times that of the Sun, is particularly intriguing not only because of its dramatic demise but also due to its age. The light from this fading dot has traversed vast cosmic distances and endured an extensive journey through time. Observing the afterglow of this cosmic explosion allows astronomers to witness an event that occurred over thirteen billion years ago, just around 600 million years after the birth of the Universe.

As of the filming of "Wonders of the Universe" in autumn 2010, GRB 090423 held the title of the oldest single object ever observed. However, it's worth noting that subsequent to the filming, a galaxy named UDFy-38135539 was discovered in the Hubble Space Telescope's Ultra Deep Field Image, slightly surpassing GRB 090423 in age. UDFy-38135539, with a light travel time of slightly over 13 billion years, currently holds the distance and age record, considering the expansion of the Universe. Despite this adjustment, the discovery of GRB 090423 remains a remarkable glimpse into the grand timescales of the early Universe, offering insights into the explosive demise of one of its first stars.

✛ STAR'S DESTINY

At present, we find ourselves amid the Stelliferous Era, a remarkable epoch that has unfolded over the course of 13.7 billion years since the inception of the Universe. This era is marked by the intricate interplay between gravity and nuclear fusion, generating a dynamic and ever-changing celestial tableau. Despite the seeming gradual pace to human observers within the span of a century, it is essential to recognize that we are currently experiencing the most vibrant and productive phase that the universe will ever undergo.

During this Stelliferous Era, the cosmic narrative unfolds with the birth and demise of stars, such as the captivating tale of GRB 090423, playing out across the night sky. These celestial bodies, though appearing timeless from our fleeting human perspective, possess destinies as certain and defined as our own. This universal drama encompasses not only distant stars but also our very own solar system's central luminary—the Sun.

The Sun, a celestial entity that has been an ever-present fixture in the sky for humanity, was formed approximately 4.57 billion years ago. Emerging from the collapse of a vast cloud comprised primarily of hydrogen and helium, with a sprinkling of heavier elements, the Sun has been a constant in the lives of humans, marking the passage of days as it rises and sets. However, it was only in the twentieth century that scientific understanding revealed a profound truth—the fires of the Sun, which have fuelled its brilliance, are not eternal. One day, in the vast cosmic timeline, the radiance of our Sun will inevitably dim, marking another chapter in the grand narrative of the Universe.

✦ DEMISE OF UNIVERSE

The Sun, our celestial life-giver, currently resides in the middle phase of its stellar journey. Through the awe-inspiring process of nuclear fusion, it is

converting vast amounts of hydrogen into helium at a staggering rate of around 600 million tonnes every second. This cosmic furnace has sustained its luminosity and warmth, nurturing the grandeur and beauty within its cosmic empire.

This majestic display of stellar energy production is anticipated to persist for another five billion years. However, as the Sun progresses through its life cycle, a momentous transformation awaits—a transformation befitting the magnificence it has bestowed upon its cosmic realm. In the final chapters of its existence, the Sun will not simply fade away. As the hydrogen fuel in its core becomes depleted, the Sun's core will undergo a collapse. In this transformative process, helium will begin to fuse into heavier elements like oxygen and carbon.

During this cosmic crescendo, a last burst of energy will be released, causing the Sun's outer layers to expand. Initially imperceptible, this expansion will manifest as an increase in the Sun's diameter by approximately 250 times its current size. The fiery surface of our star will extend beyond its current boundaries, reaching out toward Mercury, then advancing past Venus, and ultimately enveloping our delicate world. This phase, known as the red giant phase, will mark a celestial spectacle, transforming the familiar solar landscape into a cosmic theatre of colossal proportions.

The consequences for our world will undoubtedly be as disastrous as they are certain. The Earth will warm up gradually. The last perfect day on Earth will come for someone in the far future, assuming any of our descendants are still here. Our seas will evaporate, the molecules in our atmosphere will be stirred up and launched into space, and the record of life on Earth will be lost to history—or maybe to none at all if we have stayed put. This is all due to the Sun's encroaching surface.

The Sun will fill the horizon long after life has vanished; it might even reach beyond Earth. The Red Giant phase, so named because of its swelling, is the culmination of a star's energy release and the start of an extended period of decline. Within six billion years, our sun will release its outer layers into space, creating a stunning display of colour and light known as a planetary nebula. We are aware of this since we have witnessed the same series of events take place on someone else's sun—in the last gasp of far-off stars. The echoes of our future are written in filamentary patches of colour across the night sky.

If, at some point in the far future, scientists on a planet that has not yet formed look through a telescope at our planetary nebula and are struck by its beauty, they could catch a softly lit ember at its centre—all that's left of a star that we once considered to be glorious. She will be a fraction of Earth's brightness, less than a millionth of Earth's current volume, and smaller than Earth overall. Nearly every star in our galaxy will have turned into a white dwarf, including our sun, which will then be a dying, dense remnant briefly hidden by a colourful cloud.

If our planet makes it through, all that will be left is a charred and desolate rock that is silhouetted against the fading remnants of a star.

The brightest star in our sky, Sirius, is one of our closest neighbours at a distance of slightly over eight light years. It is sometimes visible during brilliant twilight because to its brightness, which is caused by both its close closeness and the fact that it is 25 times brighter and twice as large as our sun. Thus, it should come as no surprise that the earliest astronomical records contain observations of Sirius.

For millennia, humanity gazed upon the night sky, perceiving Sirius as a singular radiant star.

However, in 1862, the meticulous observations of American astronomer Alvan Graham Clark unravelled a hidden secret concealed within Sirius's luminous glow—a companion, Sirius B, lurking in the glare of its brilliant counterpart.

The Hubble Space Telescope, with its precision and clarity, captured an image (bottom right) revealing Sirius B as a faint dot of light, nestled in the lower left-hand corner. Despite its dimness, Sirius B stands as a testament to the wonders of stellar evolution. This white dwarf star, one of the larger ones discovered, possesses a mass akin to our sun, densely packed into a sphere the size of Earth.

White dwarfs like Sirius B are celestial remnants, devoid of fuel for further fusion. Instead, they radiate a faint glow, sustained by the lingering heat from their extinguished furnaces. Comprising primarily of oxygen and carbon, the byproducts of helium fusion, these white dwarfs exhibit a density a million times greater than their younger, active stellar counterparts.

In contemplating the fate of our own star, the Sun, this cosmic revelation provides a glimpse into the future. An estimated 6 billion years from now, the

Sun will undergo a similar transformation, transitioning into the white dwarf phase. As it enters this final chapter, the Sun will gradually cool in the frigid expanse of deep space. The once radiant Sun, responsible for sustaining life on Earth, will diminish to the luminosity of a full moon on a clear night, casting its ethereal glow from afar. This cosmic metamorphosis, witnessed through the lens of Sirius B, reflects the inexorable journey of stars toward their ultimate destinies.

Every star must meet death. The universe will eventually enter an endless state of darkness when all of the night sky's lights eventually disappear. The most significant result of time's arrow is that the universe, which is structured and contains all of its wonders—galaxies, planets, and stars—is finite. Many billions of stars will come and go as we proceed through the aeons that make up the age of stars. Ultimately, though, there will only be one kind of star left to light up the universe as it ages.

⊥ DEATH OF THE SUN

As we delve into the cosmic chronicle of our celestial companion, the Sun, we confront the inevitable narrative of its ultimate fate—an astronomical saga that unfolds over billions of years. Born 4.57 billion years ago from the gravitational embrace of a collapsing cloud of hydrogen and helium, the Sun embarked on a journey that has shaped the fabric of our solar system.

Currently residing in the zenith of its existence, the Sun engages in the relentless alchemy of nuclear fusion, converting hydrogen into helium at a staggering rate of around 600 million tonnes per second. This exuberant dance of fusion sustains the Sun's radiant glow, a spectacle that bathes our solar system in life-giving warmth. Yet, even in the cosmic ballet of stars, time is an unwavering choreographer, guiding each luminary through distinct stages of existence.

The Sun, now in the throes of its middle age, basks in the pinnacle of its luminosity. However, this phase is ephemeral, lasting a mere five billion more years. Despite the grandeur it exudes, the Sun, like all celestial entities, is subject to the cosmic laws of transformation.

As the celestial clock ticks away, the Sun's vast reservoirs of hydrogen gradually wane. In this inexorable cosmic narrative, the Sun's core faces depletion, leading to a cataclysmic metamorphosis. The impending fate involves the core's collapse, initiating a sequence where helium fuses into oxygen and carbon. In a climactic release of energy, the outer layers of the Sun expand, propelling its diameter to increase by approximately 250 times.

This cosmic spectacle unfolds imperceptibly, yet its repercussions cascade through our solar system. The fiery envelope of the Sun extends its reach beyond the confines of Mercury, enveloping Venus and progressing towards Earth. This celestial choreography paints a tableau of both beauty and destruction—a narrative of celestial proportions.

As the Sun undergoes this stellar metamorphosis, Earth witnesses the gradual transformation of its solar benefactor. The once-distant orb in our sky's metamorphoses into an expanding

sphere, radiating heat and light in a transformative cosmic ballet. The Sun's destiny, intricately entwined with the fate of its orbiting planets, unveils a chapter where the balance of gravitational forces and nuclear fusion yields to the inexorable passage of time.

In contemplating the demise of our solar beacon, we confront the profound interconnectedness of cosmic forces and the ephemeral nature of celestial majesty. The Sun's luminosity, once deemed eternal, succumbs to the relentless march of time. This cosmic narrative, intricately woven into the fabric of the universe, portrays the Sun's death as both a denouement and a prelude—a celestial ballet that transcends individual lifetimes, echoing through the vast expanses of time and space.

✦ LAST STARS

Proxima Centauri is the star that is closest to our solar system. Despite being only 4.2 light years away, Proxima Centauri cannot be seen with the unaided eye from Earth and doesn't even stand out in most photos that have been taken of it due to its distance from more distant stars. Because Proxima Centauri has only 12% of the mass of our solar, it is incredibly small in comparison to it. As a result, it appears to us that it shines 18,000 times fainter than our sun.

Red dwarf stars, like Proxima Centauri, are the most prevalent kind of stars in the cosmos. Red dwarfs are small and chilly, with surface temperatures around 4,000K, but they do have one benefit over their more brilliant and magnificent stellar siblings: due to their small size, red dwarfs burn nuclear fuel very slowly, which allows them to live for trillions of years. This implies that the last stars in the universe to exist are those like Proxima Centauri.

It's conceivable that our distant ancestors will centre their civilizations around red dwarfs in order to harness the energy of those final, fading stars if we do, in fact, survive into the far future of the universe. Humans may eventually get their heat from a red dwarf as the final remaining source of energy in the universe, just as our predecessors gathered around campfires on chilly winter nights.

These red dwarfs are able to live longer because the rate of fusion reactions in their cores is very low, which provides the heat pressure needed to withstand the inward pull of their weak gravity.

These stars are still active, though, and the turbulent convective currents that continuously churn their interiors cause turbulence on their surfaces. In the middle of all this activity, there are nearly constant explosive solar flares that shoot X-rays and flashes of light into space.

But in the end, these stars' thrift is no protection against the arrow of time. Proxima Centauri's fuel stores will eventually run out in four trillion years, or 300 times the age of the universe today, at which point the star would gradually collapse into a white dwarf. The universe will eventually end after trillions of years of star birth and death, leaving only white dwarfs and black holes. This period of the cosmos' history will end in around 100 trillion years, at which point it will transition into the Degenerate Era. Nevertheless, the great majority of the history of the universe is still to come—even after 100 trillion years of light. Vacant, lifeless, and bleak, the universe will continue as it enters the dark.

✦ BEGINNING OF THE END

The Skeleton Coast, nestled along the northern coast of Namibia, stands as an ominous convergence of nature's extremes—the frigid waters of the South Atlantic colliding with the arid expanse of the Namib Desert. Renowned as one of the most unwelcoming places on Earth, this desolate stretch has invoked fear since the era of seafaring mariners. In the annals of maritime history, seventeenth-century Portuguese sailors christened it 'the gates to hell,' while the indigenous Namib Bushmen, with an inherent connection to the land, dubbed it 'the land God made anger in.'

Modern explorers, armed with sturdy 4x4 vehicles or arriving effortlessly by helicopter from the port city of Walvis Bay, can now access the Skeleton Coast. However, even with contemporary means of transportation, the gods' purported anger lingers. As one stands on the sandy expanse beside the South Atlantic, the atmosphere is tinged with an ethereal eeriness.

Each morning, a dense ocean fog unfurls along the coastline, a product of the cold Benguela current's upwelling. This atmospheric phenomenon, coupled with the ever-shifting sandbanks sculpted by the relentless winds from the Atlantic, forms a treacherous alliance against navigators. Over the centuries, this toxic conspiracy has led to the demise of thousands of ships along the Skeleton Coast. The decaying remnants of rusting vessels and the bleached bones of marine life, swept ashore by the relentless currents, contribute to the coastal landscape's gothic ambiance.

The very name, Skeleton Coast, bears witness to the numerous lives lost in this desolate realm. Even for those fortunate enough to survive a shipwreck and make it ashore, the onshore currents present an insurmountable barrier. With no possibility of rowing back to sea, the only escape route leads through hundreds of miles of inhospitable desert. This forsaken stretch of land truly lives up to its reputation as a place of no return—once shipwrecked here, it marked the end of one's universe, a testament to the unforgiving nature of the Skeleton Coast.

The 91-meter (300-foot), 2,272-ton steamship Eduard Bohlen, which was traveling from Germany to West Africa on September 5, 1909, was one of the ships that met her demise here. She has been pushed hundreds of meters inland by a century's worth of shifting sands, and the Atlantic winds have ravaged her corpse, leaving it skeleton and rusting. She is surrounded by a phalanx of jackals when we get there, but they are not as afraid of us as I had thought. She provides an ethereal background for our tale; the symbolism is clear-cut, even cruel, and, in my opinion, surprisingly potent. These debris fields, intricate constructions demolished over time, resemble our dying stars.

The final surviving stars in the cosmos will not be able to defy the second law of thermodynamics any more than Eduard Bohlen did. As the universe is methodically disassembled by the laws of physics, even the white dwarfs must eventually disappear. The dying stars will gradually stop producing visible light as their heated remnants escape into space. The last beacons shining in the cosmic sky will eventually grow cold and dark after trillions of years; their leftovers are known as black dwarfs.

Black dwarfs, enigmatic entities in the cosmic tapestry, represent the culmination of stellar evolution—an ultimate fate tied to the relentless march of time. These dark, dense remnants, born from the ashes of long-extinguished stars, are currently absent in the universe due to the extended timescales required for their formation. Remarkably, despite the absence of direct observation, our grasp of fundamental physics empowers us to formulate concrete predictions regarding their destiny.

Drawing an analogy to the iron structures weathered by the winds along the Skeleton Coast, the eventual fate of black dwarfs is envisioned as a gradual dissipation into cosmic oblivion. The matter encapsulated within these celestial remnants, considered the last vestiges of material in the universe, is theorized to undergo evaporation. Over unimaginable spans of time, this matter transforms into radiant energy, drifting away into the vast emptiness of space.

The intricate processes governing the hypothetical decay of matter over such timescales remain elusive to our current understanding of physics. A more sophisticated framework, a Grand Unified Theory, is postulated to unravel the mysteries of forces at play within protons, neutrons, and electrons over trillion-year epochs. The quest for certainty in the behaviour of sub-atomic particles drives ongoing experiments worldwide, with a particular focus on probing the lifetime of protons.

Despite the absence of concrete observations of proton decay, speculation emerges, offering a plausible narrative for the eventual conclusion of our universe. The prospect of a Grand Unified Theory suggests the existence of mechanisms enabling even the most stable sub-atomic particles to undergo decay, releasing radiation as a final cosmic exhalation. While the journey into the speculative realm invites uncertainty, this envisioned tale aligns with the current contours of our understanding of physics, presenting a probable scenario for the ultimate fate of the cosmos.

In the twilight of cosmic existence, the remnants of once-majestic black dwarfs will have dissipated into the vast cosmic void, leaving behind a universe devoid of any atoms of matter. A profound transition unfolds, marking the epoch where only particles of light and enigmatic black holes persist in the cosmic tableau. Yet, even these formidable black holes, cosmic behemoths sculpted by the death throes of massive stars, are destined to undergo a process of evaporation over inconceivable stretches of time.

In the distant future, envisaged after an unfathomable expanse, the last vestiges of black holes are projected to fade away, culminating in a universe transformed into an expansive sea of light. Photons, the elementary particles of light, will dominate this ethereal expanse, converging towards a uniform temperature as the universe continues its inexorable expansion, progressively cooling these photons towards absolute zero.

Attempting to grasp the enormity of this temporal expanse, we encounter a staggering figure: ten thousand trillion trillion trillion trillion trillion trillion trillion trillion years. Represented in scientific notation as 10^{100} years, this colossal number surpasses the bounds of human comprehension. To illustrate its magnitude, envision a hypothetical scenario of counting, with each atom representing a single year. The sheer enormity of this number becomes apparent when considering that there wouldn't be enough atoms in the entirety of stars, planets, and galaxies within the observable universe to make a significant dent in this astronomical timescale. The universe, now reduced to a cosmic expanse of light, silently unfolds into a future that transcends the limits of human imagination.

The narrative of our universe will eventually come to an end when the last remaining stars have completely disintegrated and everything reaches the same temperature. It will be the first time in the Universe's existence that everything is fixed and unchanging. Because the universe is incapable of becoming any more disorganized, entropy eventually stops growing. Nothing occurs, and nothing continues to occur indefinitely.

The cosmos will stay huge, frigid, desolate, and unchanged for the remainder of time during this period, which is known as the "heat death of the universe." Because nothing in the universe

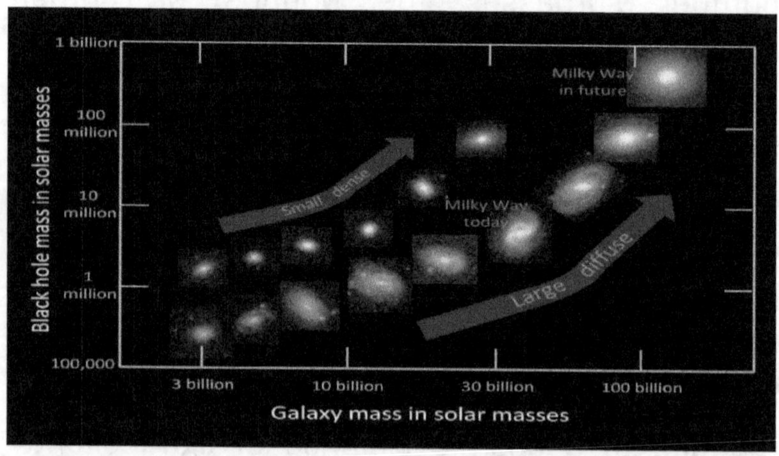

changes, it is impossible to measure how much time has passed. Since there are no temperature variations, there is no means to move energy in order to cause anything to happen. As a result, nothing changes. Time's arrow has just stopped existing. Because it is a requirement of the fundamental principles of physics, this reality cannot be avoided. The universe is destined to end; all the hundreds of billions of stars within the hundreds of billions of galaxies will burn up and take with them the chance for life to exist anywhere in the cosmos.

✛ VERY PRECIOUS TIME

It may seem a little bleak to you that the Sun will eventually die, destroying Earth and all life on it, and that the other stars in the Universe will do the same, leaving behind a vast, formless cosmos that will never be able to support life of any kind or retain any trace of the living things that gave its past meaning. It is reasonable for you to have concerns regarding the construction of our universe. You could surely construct a universe in a different way, right? You must have created the universe so that order didn't have to give way to chaos, right? The short answer is that, if you wanted life to exist in it, you couldn't.

The exact thing that initially made the conditions for life possible is the arrow of time, the series of alterations that will gradually but inescapably bring the universe to an end. It took time for matter to develop in the Universe after the Big Bang and for it to cool down sufficiently; it also took time for gravity to bring matter together to form planets, stars, and galaxies; and it took time for matter on Earth to organize into the intricate patterns known as life. Each of these actions was compliant with the Second Law of Thermodynamics and represents a step along the lengthy path from order to disorder.

Though it won't last long, the arrow of time has produced a brief, dazzling window in the Universe's adolescence when life is feasible. Life on Earth is only feasible for a billionth of a billionth, billionth, billionth, and so on, as a percentage of the universe's lifespan, which is calculated from the beginning of time until the evaporation of the last black hole. Because of this, I believe that a single instant in time rather than a star, planet, or galaxy is the most amazing wonder of the universe. And the moment is here.

3.8 billion years ago, life first appeared on Earth; 200,000 years ago, the earliest humans roamed the African plains; 2.5 million years ago, people thought the Sun was a god and used stone towers perched atop hills to calculate its orbit. These days, science rather than solar gods satisfies our curiosity, and we have observatories that can peer deep into the universe and are nearly exponentially more advanced than the Thirteen Towers. We now know a great deal about its present because we have seen its past. Even more astonishingly, we are able to compute and make specific predictions about the future state of the Universe by employing the complementary fields of theoretical physics and mathematics.

I think the only ways we can truly comprehend ourselves and our place in this amazing universe are by gazing up at the sky, carrying on with our study of the universe and its laws, and letting our curiosity run wild as we explore the boundless natural world.

Voyager 1 was a spacecraft that was launched on a "grand tour" of the Solar System in 1977. Before leaving for interstellar space, it made amazing findings while visiting the big gas giant planets Jupiter and Saturn. After nearly finishing its mission thirteen years later, Voyager turned its cameras around and snapped a final image of its house. The Pale Blue Dot is the image on the left. The single point of light at the centre is the most beautiful item ever photographed—possibly even the most beautiful thing ever seen—because it represents our planet, Earth. Having been taken more than six billion kilometres (3.7 billion miles) away, this is the furthest image of our planet ever captured.

The striking and profound aspect of this little point of light is that all known life in the history of the universe has existed on that pixel—a pale blue dot suspended against the void of space—for some portion of its existence.

It has been argued that studying astronomy is a humbling and character-building experience, as noted astronomer Carl Sagan once wrote. Perhaps there is no greater illustration of the foolishness of human conceit than this far-off picture of our small planet. It emphasizes, in my opinion, how important it is that we treat each other with more kindness and protect the little blue dot that has been our sole home.

Life in the Universe will only last for a brief, brilliant moment in an endless amount of time, just as we, and all life on Earth, stand on this small speck drifting in vast space. This is because, like the stars, life is a transitory structure on the long journey from order to disorder.

Even so, life is the way the universe, if only momentarily, can comprehend itself, which does not make us unimportant. In the short time we have lived on Earth, we have sent spacecraft to the farthest reaches of our solar system and beyond, constructed telescopes that allow us to view the oldest and farthest stars, and identified and comprehended at least some of the basic laws that govern the universe. In the end, I think this is the main reason we matter. Our genuine importance is in our unwavering quest to comprehend and investigate this amazing, exquisite, yet transitory home—the universe.

ABOUT THE AUTHOR

The visionary Founder & CEO of **MB WEBBER'S**, a state-of-the-art Software Development company situated in Manbazar, Purulia, West Bengal, is **Pabitra Banerjee**, an innovative **Full-Stack AI Engineer**. Pabitra, who was born on **January 22, 2004**, has grown to prominence in the technological industry quite rapidly.

He demonstrates a unique combination of technological expertise and a love of mathematics and astrophysics. Pabitra Banerjee, the sole child of **Mr. Pulak Banerjee** and **Mrs. Babita Banerjee**, is committed to promoting knowledge and comprehension of science and technology. His efforts go beyond his workplace; he is actively involved in cooperative learning settings with organizations such as **Code Explorer** and **Dev Line Community**.

As a well-known author in the field of astrophysics, Pabitra regularly offers opinion on the most recent space missions carried out by international space agencies. In **MARVELS OF THE COSMOS**, he steps into the genre of prose while combining his technical expertise with a desire to provide readers with the information they need to successfully traverse the knowledge about our universe.